FACILITY MANAGEMENT SYSTEMS

FACILITY MANAGEMENT SYSTEMS

JEFFREY M. HAMER

VNR VAN NOSTRAND REINHOLD
New York

For Deborah, Stephen, Jonathan, and ?

Copyright © 1988 by Van Nostrand Reinhold
Library of Congress Catalog Card Number 87-29570
ISBN 0-442-23245-4

Printed in the United States of America
Designed by Sharen DuGoff Egana

Van Nostrand Reinhold
115 Fifth Avenue
New York, New York 10003

Van Nostrand Reinhold International Company Limited
11 New Fetter Lane
London EC4P 4EE, England

Van Nostrand Reinhold
480 La Trobe Street
Melbourne, Victoria 3000, Australia

Nelson Canada
1120 Birchmount Road
Scarborough, Ontario, M1K 5G4, Canada

16 15 14 13 12 11 10 9 8 7 6 5 4 3 2

Library of Congress Cataloging-in-Publication Data
Hamer, Jeffrey M., 1949–
    Facility management systems.

    Bibliography: p.
    Includes index.
    1. Facility management—Data processing.   I. Title.
TS177.H36   725'.4'0285   87-29570
ISBN 0-442-23245-4

CONTENTS

PREFACE

Fundamental forces are changing our economy. The stock markets have experienced highs and lows that would have been unthinkable only a few short years ago. Competition that was national is now global. Struggling to become more productive in their newly expanded competitive arenas, corporations, institutions, and government are restructuring by means of asset revaluation; that is, they are seeking to have maximum value placed on their assets by the marketplace and investors.

In its publication "The Changing Economy: Evolution in Perspective," the investment banking and brokerage firm of Drexel Burnham Lambert makes the following observations: "[American corporations are] looking for ways to increase profits and productivity and fully utilize hidden assets . . . . Corporate leaders are taking a closer look at their existing assets . . . . The competitive vitality of the U.S. depends on rearranging the balance sheet." Thus, managers and corporate raiders alike have sought to raise the value and productivity of the asset side of balance sheets. As Robert Anderson, Chairman of Atlantic Richfield Company, states, "If you can't manage assets to reflect their true value, you're inviting someone to do it for you."

Facilities are the largest single class of asset on the balance sheet. Managers have a new awareness of the strategic opportunity presented by facility management. A significant part of this opportunity is facility management *systems:* the information tools that enable managers to make informed facilities decisions.

This book is intended primarily for readers who deal with "real world" facility management problems: people who require information about the what, when, and where of space, furniture, equipment, occupancies, moves, leases, and the like. Included in this audience are people in management who require "bottom-line" information for decision making, as well as facility engineers and managers who must implement decisions and respond to occupants' needs on a daily basis. Practitioners such as architects, engineers, interior designers, and environmental designers also will find information here that is of significant interest.

While a certain amount of general and theoretical background is desirable or even mandatory in order to apply the tools effectively, most of the material is intended to be practical. Significant time is spent on the subject and practice of dispassionate cost/benefit analysis, for example.

Computers are playing an increasingly important role in the design and ongoing management of buildings. In the near future, I believe this role will be central. Thus, this book focuses largely (but by no means exclusively) on computer-based and computer-assisted techniques.

I and my associates strongly believe, however, that most of the benefits which organizations achieve through automation in facility design and management are obtainable without a computer. These benefits come from the required logical preorganization of process that the highly structured nature of today's computers require. Thus, a computer is not essential for gaining many of the advantages that flow from computerization. This book focuses on computer techniques and human procedures, not on computers.

ACKNOWLEDGMENTS

This book could not have been prepared without the advice, review, contribution, and support of my partners Bill Mitchell, Ched Reeder, Tom Kvan, and Eric Schreuder. Substantial contribution was made by Jim Steinmann. The work of the entire staff of The Computer-Aided Design Group significantly affected the content.

Important editing and organizational work was done by Audry Fisher of Michael Klepper and Associates. Jean Fallowfield assisted with editing and compiling the manuscript. Our associate Robin Liggett also provided material. Rhonda Curtis assisted greatly with final production. Paula Shaffer and Arlie Fitch assisted with word processing. Bill Kovacs provided advice on and production of the cover illustration.

I am grateful to the following individuals for sharing their thoughts, work, or both: John Adams, Chuck Atwood, Howard Burger, Basil Calliamannis, David Culbreth, Malcolm Davies, Ron Delp, Richard Dilday, Ellen Dupps, Charles Eastman, Anne Fallucchi, Don Fullenwider, Ed Forrest, Joe Fox, Natalie Gerardi, John Grower, Lee Hales, Mike Hamman, Jere Hunter, Ken Johnson, Warren Joiner, Peter Kimmell, Warren Koepf, Ron Kramer, Adora Ludy, Allen Lungo, Chris Miller, Tony Mirante, Stanleigh Morris, Virginia Mullen, David Necker, Mary Oliverson, Wayne Pierce, Terry Poindexter, Ed Popko, Jack Robinson, Ed Rondeau, Mike Schley, Mel Schlitt, Larry Schnur, Adrian Schreiber, Ann Seltz, Gary Silver, Sheri Singer, Robert Sison, Terry Stretch, Mike Tatum, Lou Thomas, Linda Tasker, Janice Tuchman, and Janet Waxman.

I also gratefully acknowledge contributions and information from the following organizations: Auto-trol Technology; Welton Becket Associates; Everett I. Brown Company; CADAM, Inc.; California Computer Products, Inc.; Carnegie-Mellon University; CHA; ComputerVision, Inc.; CRS/Sirrine, Inc.; DFI/Systems; Digital Equipment Corporation; Duffy, Inc.; Equitable Real Estate, Inc.; Facility Design & Management; Facility Management Consultants; Facility Management Institute; Facility Programmatics, Inc.; Formative Technologies; Fullenwider Consulting Group; GE Calma Co.; Gensler & Associates; Harvard Real Estate, Inc.; Hellmuth, Obata & Kassabaum, Inc.; Herman Miller, Inc.; Intergraph Corporation; International Business Machines Corporation; International Facility Management Association; Albert C. Martin & Associates; Massachusetts Institute of Technology; Mc-Donnell-Douglas Corporation; McGraw-Hill, Inc.; Micro-Vector, Inc.; Price Waterhouse; Real

Estate Research Corporation; Skidmore, Owings & Merrill; Steelcase, Inc.; Steinmann, Grayson, Smylie; Swimmer, Cole, Martinez, Curtis; the United States General Services Administration; the University of California at Los Angeles Graduate School of Architecture and Urban Planning; and Warefront Technologies.

The indulgence and support of my remarkable wife and family enable me not just to write a book, but to enjoy life in general.

ACKNOWLEDGMENTS

x

INTRODUCTION

In recent years, the areas of *facility management* and *space management* have elicited great interest on the part of real-estate groups and facilities engineering departments of major corporations and institutions, the government, and the environmental design disciplines. Facility management has been promoted as the solution to most problems encountered by building occupants, owners, managers, and even architectural designers. Although the reality of today's facility management techniques and tools falls significantly short of such status, they will significantly help a majority of building owners and occupants.

There is more space within existing buildings than there is new space being built. The corporate or institutional space manager can find more opportunities for satisfying space needs in existing space than in new space—quite apart from consideration of the resulting cost savings. Similarly, there is potentially more work for the design practitioner in optimizing the use of existing space than in focusing primarily on new construction.

While much lip service has been paid to this theme for more than a decade, events only recently have made facility management systems both desirable and practical.

The practice of facility management as a professional discipline (with its own defined procedures, professional and educational associations, and the like) has begun. The growth of the most prominent trade organization in this field—The International Facility Management Association (IFMA)—from its inception in 1980 to 1,500 members in 1985 and more than 4,000 members in 1987 is one dramatic example of this. More than half of the respondants to IFMA's 1986 survey had staffs of sixteen to twenty people. This same research (*IFMA Report Two*, 1986) documents "a significant trend toward doing more functions with in-house staff." More than two-thirds of these individuals are using computers—most for the first time, as this research shows an increase of greater than 400 percent in computer use over the previous year.

Macroeconomic trends indicate that facilities costs constitute a growing portion of operating costs (fig. 1). Meanwhile, the cost of automation has declined radically (fig. 2). Thus, the net effect of these two trends is an improvement in the benefit-to-cost ratio of automation tools in facility management over time (fig. 3).

Recognition by businesses of this increased benefit-to-cost relationship is evident in the fact that a greater number of facility management tasks (gathering, organizing, and presenting in-

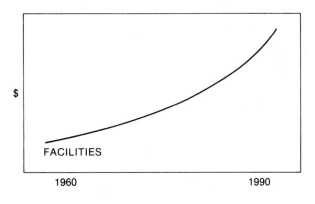

Figure 1. Trend of facilities costs over time. (Courtesy of Computer-Aided Design Group)

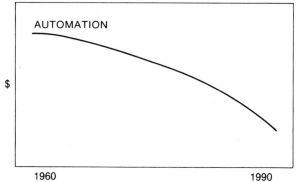

Figure 3. Trend of facility management benefits realizable through automation. (Courtesy of Computer-Aided Design Group)

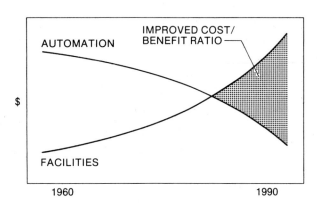

Figure 2. Trend of automation costs over time. (Courtesy of Computer-Aided Design Group)

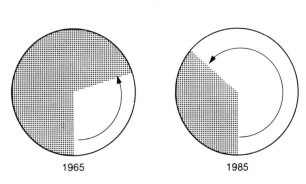

Figure 4. Percentage of facility management tasks worth automating. (Courtesy of Computer-Aided Design Group)

formation, communicating with co-workers and vendors, solving design problems, and so forth) now are candidates for automation (fig. 4). The recent recognition of increasing facilities costs has caused design practitioners, owners, occupants, and managers to reuse space in a continuous and large-scale way and to identify it as a resource of significant value.

Long lead times are needed to acquire space,

and the cost of operating in an ineffectively planned space is inordinately high—particularly for larger organizations. But this is a relatively new perception of both designers and managers; and perhaps as a result, the development of even simple procedural tools has lagged.

While dozens of organizations have rushed to market products they tout as comprehensive facility management systems, sophisticated space managers and users are virtually unanimous in feeling that their needs remain partially (and sometimes largely) unmet. Design professionals have fared somewhat better on the matter of tool availability, although the transfer of information from the designer to the ongoing occupant remains poorly understood and even more poorly addressed by products. The general availability of powerful and inexpensive computational capacity, however, is aiding the development of tools that would have been impractical even a few years earlier.

This book catalogs a representative sample of current methods and tools and provides a practical introduction to use of the tools and to means of evaluating new ones.

When we mean to build,
We first survey the plot, then draw the model;
And when we see the figure of the house,
Then must we rate the cost of the erection;
Which if we find outweighs ability,
What do we then but draw anew the model
In fewer offices, or at least desist
To build at all?

William Shakespeare
Henry IV, Part 2, I, iii

ONE

A FACILITY MANAGEMENT OVERVIEW

DEFINING FACILITY MANAGEMENT

As the field of facility management assumes greater importance and as more individuals and organizations become involved, its attendant definitions and descriptions continue to increase. One of the more straightforward of these comes from the International Facility Management Association (IFMA), which defines facility management as "the practice of coordinating the people and the work of an organization into the physical workplace." A more detailed definition is offered by *Engineering News-Record* (April 4, 1985):

> the discipline of planning, designing, constructing, and managing space—in every type of structure from office buildings to process plants. It involves developing corporate facilities policy, long-range forecasts, real estate, space inventories, projects (through design, construction, and renovation), building operation and maintenance plans, and furniture and equipment inventories.

Jim Steinmann has defined the field as "the systematic method of inventorying, planning, designing, allocating, and maintaining space, equipment, and furniture for general or special purpose facilities that are subject to a need to be flexible to accommodate change." As he points out, the greater the degree of flexibility required or the more indeterminant the future requirements, the greater the need becomes for an effective management system. The potential for long-term cost savings increases with the operation's scale and organizational complexity.

Steinmann then goes on to suggest the following facility management goals:

- Replace reaction with action.
- Plan for the future.
- Improve space utilization and efficiency.
- Promote user involvement.
- Improve productivity.
- Maximize return on capital investment.
- Minimize present-value life-cycle cost.
- Provide qualitative support of the enterprise/business process.
- Satisfy users.

The differing definitions of facility management show that it is an evolving field whose nature is still somewhat fluid. The Library of Congress seeks to encompass the field's evolution and change by defining facility management as "the practice of coordinating the physical workplace with the people and the work of the organization, integrating the principles

1

of business administration, architecture and behavioral and engineering sciences." In this book, *facility management* is defined as the process of planning, implementing, maintaining, and accounting for appropriate physical spaces and services for an organization, while simultaneously seeking to reduce the associated total cost.

Infrastructure Management and Other Definitions

This book is concerned primarily with buildings. Included within its scope are the services and infrastructures directly required to support buildings, interior space, and associated equipment.

Another kind of facility management focuses exclusively on the supporting framework of various structures. Ed Forrest has observed that this type of facility management is "typically the province of state Departments of Transportation, local public works departments; power, gas, water, sewerage and telecommunications providing organizations; typically documented on *base maps* of the community." The concern of this type of facility management is sometimes referred to in the trade press as "intelligent infrastructure." It represents a mapping (graphical database) problem, and IBM Corporation refers to it as *geofacilities* information.

Forrest continues, "Both the '*intra-structure*' facility management database and the *infrastructure* facility management database are founded on maps. Isn't a floor plan of the office building a map of sorts? Would you venture into the Pentagon without one?"

Facility management has other meanings to other constituencies. To the data-processing community, it means operation of computer rooms on a contract basis. To others, including those in the oil industry, some utilities, and branches of the military, *facility management* (or more recently, *infrastructure management*) means mapping: thus another term used in the trade press, *AM/FM*, stands for Automated Mapping and Facility Management. To still others, *facility management* refers primarily to facility operations, maintenance, or both.

Facility Management Systems

A facility management system is a product and/or process that aids in the performance of facility management functions. A facility management system generally includes each of the following (listed in order of importance):

- Procedures
- Required data
- Education and training
- Automation

This book stresses that automation is but a part—albeit an important one—of a facility management system. Procedures, information, and education, however, are even more important.

Still, there is a natural tendency in practice to focus on automation. Automation, properly designed and implemented, can incorporate procedures, information, and even education. The operative component of automation in this context is the software: the computer programs that instruct the physical devices (hardware) to process the facility management information.

Architectural Technology, a publication of the American Institute of Architects, published the following definition of *facility management software*:

Facility management software may be considered to be software that addresses many of the following types of applications: long-term planning of space requirements, development of space/furniture standards, generation of management-type reports of facilities data that relate to space utilization, reports to aid in the evaluation of the long-term facility plan, furniture and space inventory, block and stack plans, etc. Pure CAD (Computer-Aided Design), without these types of applications, is not considered to be included in CAFM (Computer-Aided Facility Management).

GROWTH OF AUTOMATED FACILITY MANAGEMENT

A computer-based systematic approach to facility management has become practical only very recently. Various situations have resulted in the dramatically increased interest in this area. First, facility management is gaining recognition as a business function and is thus becoming a high-priority computer application. Historically, top business management has identified facility management with building maintenance. Recently, however, facilities have begun to be viewed, along with people, capital, and technology, as a strategic business resource. This is because appropriate facilities resources often are very important to products and business activities moving forward; sometimes facilities resources are even the constraining or "gating" item on important business projects. The major accounting and business consulting firms are helping this "strategic resource" view of facilities to grow.

Computer resources in major organizations (hardware, software, money, and people) are allocated by priority, higher priority being given to those applications that support strategic business needs. Thus, as facility management gains recognition for its strategic importance,

its supporting computer applications are receiving higher priority.

Second, facilities costs (real estate, fixed assets, servicing costs, and support costs) have risen in real terms. The percentage of the corporate overhead "pie" represented by facilities has become larger, causing the associated costs to have a greater effect on the bottom line (fig. 1-1). The International Facility Management Association in 1984 (*IFMA Report One*) noted the following as typical annual facilities expenses:

Variable	Range
Dollars per employee	$2,250–$6,250
Dollars per square foot	$7–$11
Percent of corporate expenses	10%–18%
Percent of corporate gross income	5%–12%

A 1981 survey by Harvard Real Estate, Inc., indicated that real estate assets represent 25 percent of America's corporate balance sheet and involve 4 to 7 billion square feet of owned space and 2 to 7 billion square feet of leased space. The value of these assets was estimated at that time to be between $700 million and $1.4 trillion. Price Waterhouse subsequently estimated the value at $1.5 trillion.

Third, management demands more. Facility managers are being expected to provide more information and to report it more frequently. Anticipation and speed of reaction have become more important in real estate and fixed asset management decisions. Economic, tax, and use-of-capital issues must be addressed more rapidly than before, as facilities planning decisions become more integrated with strategic business planning.

Fourth, significant computational resources are beginning to become available to large

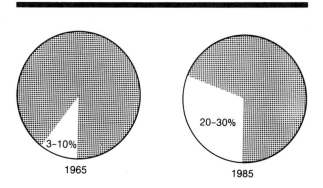

Figure 1-1. Facilities costs as a percentage of operating costs. (Courtesy of Computer-Aided Design Group)

numbers of people. Relatively inexpensive and commonly available computers offer more computational power than existed on the earth little more than twenty years ago. Most managers are learning (or have learned) to utilize their first personal computers. According to Eileen Carstairs, writing in *Corporate Design and Realty* (March 1985):

> The advent of sophisticated and affordable personal computers has been accompanied by an increased awareness by organizations of the gains to be had from better facility management and space planning. The typical organization owns property worth about a quarter of its total assets, yet many lack fundamental inventories of the actual buildings and land they own. As construction costs, rents, and mortgage rates rise, and as commercial real estate values escalate, more and more companies are looking for ways to get the most for their money.

Although this computational power is not entirely a blessing—because new tools are often used inappropriately (sometimes a personal computer is not the most appropriate tool to use in attempting to manage hundreds of buildings)—the availability of cheap computational power to facility managers is overwhelmingly positive. Computers will become more and more prevalent in management, in part because of the trend in cost of computation as a percentage of the cost of the user (see figs. 1-2 and 1-3): Computational costs are falling; computational power is rising; labor costs are rising. In sum, the cost of computation as a function of the cost of the user's labor is falling at an almost geometric rate. This is a fundamental and very important trend. Figure 1-3 shows that the annual cost of a computer workstation was 80 percent of the cost of its user in 1975. By 1985, this cost had declined to 5 percent of the cost of its user—a sixteenfold reduction. Because databases for all but trivial problems are huge, the storage, computational power, and data management requirements are

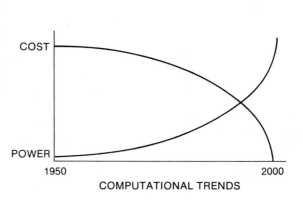

Figure 1-2. Trend lines of computer cost and computer power.

substantial. Cost-effective facility management had to wait for the reductions in computer costs that have fueled automation in many industries.

Fifth, computer tools are improving. Until quite recently, software and procedures for facility managers lacked sophistication. Ease of use was also an issue. Both technologies are advancing rapidly. For example, powerful, reliable computer-aided design and drafting (CADD) systems (fig. 1-4) are available for graphic representation of facility management decisions. Another important advance occurred in the area of data entry (the process of locating, entering, and checking information). The cost of data entry is the largest cost of any computer application. Only when information can be accessed for more than one application can its cost of entry be effectively amortized. In recent years, modular systems have begun appearing (linked to central databases) that address all areas of facility management and eliminate the need for multiple data entry.

Multiple applications of a single CADD system at workstations like the one shown in figure 1-4 leverage information entry costs still further. These CADD systems are beginning now to interface closely with facility management de-

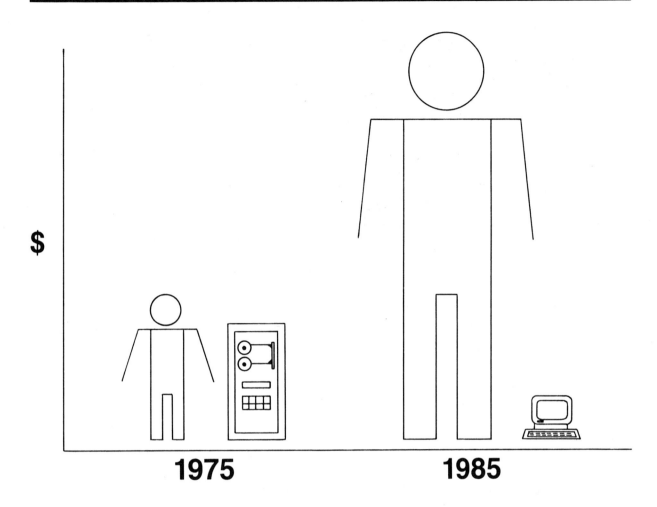

Figure 1-3. Computer cost versus user cost over time.

cision systems. Both facility management systems and CADD systems are appearing for the industry-standard mainframe and departmental computers that major space users already own. Multiple applications running on a single computer leverage investments in hardware, in training, and (most significantly) in data collection costs.

Despite the increased attention being paid to facilities management by top management per-

sonnel and despite the increased availability of competent software, major organizations only recently have begun to install and utilize computer-based facilities management systems. Less than half of the major organizations in the United States that have 500,000 square feet or more of working space have automated facility data of any kind. Of those that do, only 5 percent have systems with a "what if?" or modeling capability.

Figure 1-4. Typical CADD workstation. (Courtesy of CA-DAM, Inc.)

FACILITY MANAGEMENT ACTIVITIES

The functions implied by our definitions of facility management—and the activities usually performed by facility management professionals—are quite broad; they include the following:

• Inventory management
• Requirements programming
• Master planning
• Location and layout planning
• Drafting
• Cost accounting
• Real-estate strategy
• Move coordination
• Project administration and implementation
• Purchasing coordination
• Maintenance planning
• Site management
• Overall system coordination

These and other facility management activities will be discussed in greater detail in subsequent chapters.

The Computer-Aided Design Group has or-

ganized all facility management activities into a simpler scheme encompassing six areas: master planning, project design, project implementation, facility operation, facility inventory, and group administration (fig. 1-5).

The firm of Steinmann, Grayson, Smylie has organized similar information as shown in figure 1-6. That its functions and activities can be viewed and organized in different ways shows that facility management is still a developing field.

Personnel Involved in Facility Management

Facility management is a multidisciplinary function that generally involves more than one department in a large organization. Many professionals are involved in the decisions, although few are actually called "facility managers." Examples of such professionals include the following:

• Engineers
  • plant engineers
  • facility engineers
  • site engineers
  • maintenance engineers
• Facility administrators
  • security
  • fire, health, and safety officers
• Planners
  • strategic planners
  • energy planners
• Designers
  • architects
  • space planners
  • interior designers
• Managers
  • real-estate managers
  • administration managers
  • administrative services managers
  • planning managers
• Developers

Figure 1-7 illustrates how one company has organized its facility management personnel.

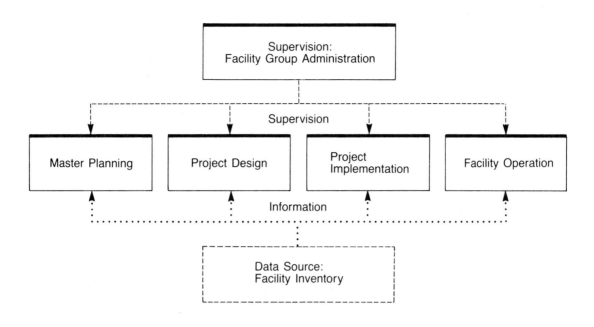

Figure 1-5. Facility management. (Courtesy of Computer-Aided Design Group)

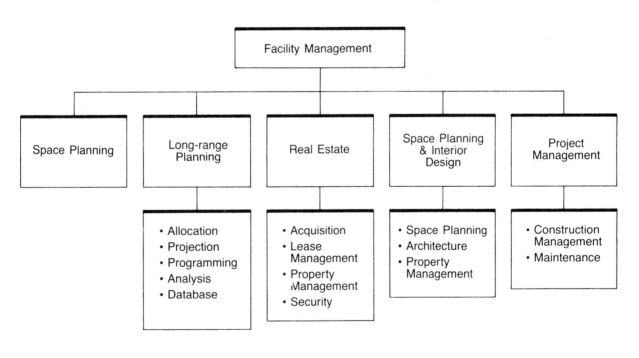

Figure 1-6. Facility management overview. (Courtesy of Steinmann, Grayson, Smylie)

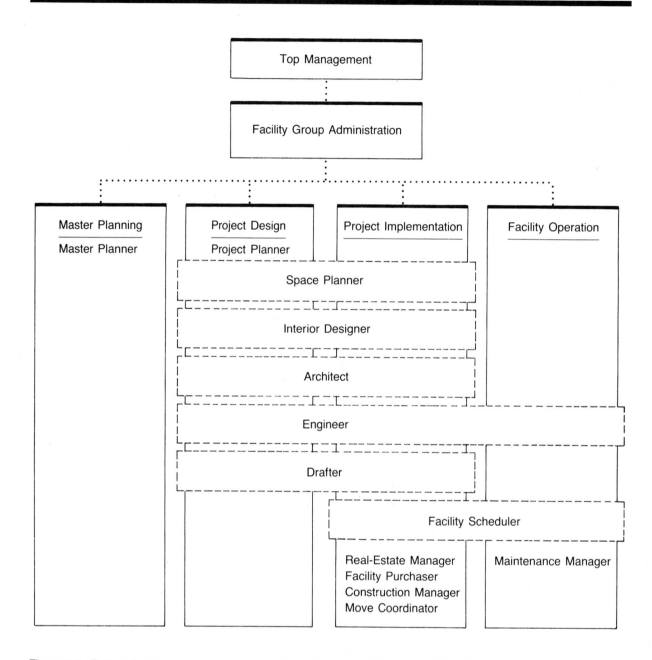

Figure 1-7. Typical facility management organization. (Courtesy of Computer-Aided Design Group)

Organizations Involved

The organizations within which facility management tasks are performed include the following:

- Corporations
  - industrial corporations
  - commercial corporations
  - service corporations
  - financial corporations

- developmental corporations
- Private institutions
  - universities
  - health care institutions
- Government
  - agencies
  - military
  - school
  - federal government
  - state government
  - local government
- Property management firms
- Property investment organizations
- Service organizations
  - design firms (architecture, interior design, engineering
  - consulting firms (management consultants, facilities consultants, space programmers, accountants
- Product vendors
  - furniture manufacturers and dealers
  - equipment manufacturers and dealers
  - CADD/computer system vendors
  - building products firms

In addition to the work of corporate, institutional, and governmental facility managers, the work of many other individuals within these organizations is directly related to the discipline of facility management.

GOALS OF FACILITY MANAGEMENT

There are many reasons for developing and maintaining a facility management program. In organizations that have grown rapidly or reorganized significantly, a primary need is simply to gain control over the present situation. This involves knowing what exists, who is using it, what purposes it serves, and how much it costs. As total facility-related costs (such as rent, construction, furnishings, equipment, moving, energy and utilities, operations, security, and maintenance) continue to escalate, the need and the payback for developing a facility management program also increase.

The goals may be summarized in two broad categories:

- Reduce total facility-related costs (and such costs as a percentage of total operating costs).
- Improve productivity and functionality of the organization (by improving the housing of the enterprise).

Benefits

As of 1984, more than 25 billion square feet of major nonresidential facilities (corporate and institutional buildings in excess of 100,000 square feet) existed in the United States. If facility management techniques were applied to even one-tenth of this space, the resulting benefits could be conservatively estimated at over $1 billion per year. Even the manager of one 100,000 square-foot building should be able to attain benefits in the range of at least $50,000 annually.

Implementing a comprehensive facility management program should not only improve equipment and space utilization efficiency, but (more important) should minimize the operational costs. Facility costs include initial development costs, operation costs, maintenance costs, moving costs, costs of furnishings, and equipment costs. Charles Reeder has noted that all building capital costs are trivial in comparison to building operating costs such as wages and maintenance over time. This fact provides even more justification for reducing operating costs via a facility management program.

By implementing a facility management program, the manager should be able to accomplish the following goals:

- Develop more meaningful and accurate forecasts of future space requirements, reducing expenditure of resources.
- Prepare more accurate future capital budgets.
- Provide a framework within which to meet established budgets more effectively.
- Improve employee morale and efficiency in pro-

portion to the degree to which workstations and an improved environment better respond to employee's needs.

Additional benefits that may be derived from the implementation of a comprehensive facility management program include the following:

- Employees are encouraged to become more anticipatory and less reactionary in their facility management decisions.
- Solutions to specific problems are developed within the context of an overall space utilization master plan.
- Space utilization efficiency improves.
- A reduction may occur in new/future space required, in "staging" space, or in space held in reserve for emergencies.
- Some implementation and construction can be postponed or avoided.
- The number of rearrangement and relocation projects is reduced.
- Information is better managed, and the inventory of space, equipment, and furnishings is controlled.
- Capital resources required to support operations are controlled and allocated more effectively.
- The overall work environment improves, and a more functional, flexible, and cost-effective facility is made possible.
- Functional standards are developed for offices, workstations, equipment, and special facilities.
- Average procurement costs are reduced.
- Interior planning, design projects, and design components become standardized.
- Necessary current and future facility requirements data are kept readily available.
- Energy consumption is reduced.
- Electrical, communication, and similar services are distributed more effectively.

Variables Affecting Benefits

Six major variables affect the number and extent of benefits realizable from a space management program:

1. *Area:* The size of the facilities involved obviously affects the degree of benefit obtainable—the larger the amount of space managed, the larger the potential benefit.

2. *Type of facility:* The type of facilities involved affects both the capital costs and the maintenance costs, and thus it affects the possible benefits. Hospitals and high-technology or laboratory facilities, for example, are more expensive and complex than similar-sized facilities for warehousing.

3. *Churn rate:* This IFMA-defined term (expressed as a percentage of total space) refers to the amount of space moved or changed annually. A churn rate of 25 percent annually, for example, indicates that the equivalent of one quarter of the total space in an organization's facilities is moved (through having employees, equipment and/or activities change location) or is changed (through remodeling or new construction projects) each year. Note that the word *equivalent* was used in this explanation. The example case of a 25 percent churn rate could be achieved, for example, if half of the organization moved every two years, or if 5 percent of the organization was rebuilt five times in a single year. In reality, of course, the churn rate represents the average of all such occurences annually. The average (mean) churn rate of IFMA organizations has remained stable at about 30 percent per year for the (last several) years in which IFMA has surveyed.

   The higher the churn rate, the greater the realizable benefits.

4. *Growth/consolidation rate:* The rate at which an organization is growing (or shrinking) and the (not always corresponding) rate at which its facilities are expanding (or contracting) affect the potential benefits of facility management. (The potential benefits of a facility management program are equally great for both growing and consolidating organizations.) Rapidly expanding organizations seldom give facility management high enough priority to avoid having constraints on their facilities negatively affect their ability to provide products or services. Organizations attempting to manage shrinkage or business contraction face unique and especially difficult problems because the funds needed to implement a facility management program are often lacking due to the economic forces that prompted the consolidation in the first place.

5. *Size of the facilities staff:* Because one of the benefits of facility management is reduced labor cost

due to higher facility management staff productivity, the size of the staff presently handling facilities functions affects the potential benefits.

6. *Management's needs:* The needs and desires of an organization's management affect the benefits of a space management program, as well. This is an important variable. Some management teams require extraordinarily detailed information at relatively frequent intervals. Others are content with broad statistics, without backup, relatively infrequently—although sometimes the latter view prevails only because the former is impossible using methods already in place. In any case, the greater the demands management makes, the larger the potential benefits are of a well-conceived space management program.

CHALLENGES OF FACILITY MANAGEMENT

If problems were not encountered in facility management, this book probably would not have been worth writing nor worth reading. Following are some common challenges facing the profession.

Lack of Definition

The biggest challenge may be the lack of clear definitions of facility management and of the facility manager's role and mission. These vary tremendously from one organization to the next, as well as over time. The field has grown slowly, draws upon many sources of information, uses many techniques, and provides information of many kinds. Attempts to create facility management systems of broad applicability must encompass this diversity. The facility manager correctly asks "how do others in my industry do it?" Organizations such as IFMA are attempting to respond in a suitably comprehensive and sophisticated manner.

Increasing Scale

The increasing scale of projects seems to be a continuing trend. As buildings and their internal organizations grow, the ability of the human designer and manager to deduce or even to comprehend important facts becomes strained. Our ability as human beings to store, index, and retrieve details about larger projects is also limited.

Large organizations occasionally make expensive and embarrassing errors. On more occasions than might be expected, companies have "lost" multiple buildings or have bought buildings through realtors from themselves. This does not mean that the executives responsible are bad managers; it does mean that, beyond a certain scale, the information cannot be maintained without exquisitely well-documented and well-executed procedures and, beyond a further scale, without automation.

Information management procedures and automation tools can be important here simply in allowing organizations to find and identify a particular piece of information they want, when they want it.

Design Complexity

Increasing scale is often accompanied by increasing design complexity. The increasing sophistication of our buildings, their systems (life support and comfort, building infrastructure, transportation, communications, emergency, and so on), and their users all contribute to the complexity of the information that must be managed. The expanded range of potential solutions to problems also complicates the manager's tasks. Sometimes the answer to the question as simple as "can those conduits accept two more wires each?" requires a significant amount of time and money to answer because of the complex nature of the facility and its equipment.

The increased complexity of our projects evidences itself also in the relationships among elements. We can now specify required and desired relationships between and among people, spaces, activities, objects, pieces of equipment, and so on, with greater accuracy and precision than ever before. As we become aware of these relationships in greater detail, we seek tools that will enable us to respond to these relationships and to satisfy their requirements. Examples are the relationships between equipment or activities that make noise and people that require quiet.

Organizational Issues

While buildings clearly have a physical form, they have an organizational one as well. Indeed, most of the architectural program information that a facilities manager, building owner/occupant, or architectural designer receives, reflects the occupant's organization(s). In a corporate or institutional reporting structure, this takes the form of the familiar and traditional organization chart or family tree. An example is shown in figure 1-8.

All occupancies exert their organizational influence and generate most of their information according to the form of their internal structures. Most of this information must be mapped into the physical structure again and again. A need for the accounting group to be near to the controller's office, for example, might translate into a need for the accounting group to be on a particular elevator bank, floor, or quadrant.

The organizational issues can be subdivided into a number of categories.

Conflicts and Inconsistencies: Increased scale and complexity naturally increase the potential for conflicts and inconsistencies. At a minimum, information management tools should help identify conflicts and inconsistencies; ideally they should suggest resolutions. Conflicts can take forms ranging from physical (for example, two occupancies assigned to the same physical space) to more subtle logical inconsistencies. An example of the latter would be a situation in which the attributes of spaces, activities, or pieces of equipment (such as security characteristics) were described differently (or in con-

flicting terms) in two different places in the database.

People: Increased complexity also evidences itself in personnel. Our society encourages (and our technology is beginning to permit) individuals to exert greater influence over their working lives in areas such as the working environment and work scheduling. Legal requirements are emerging for such things as ergonomics in workstations, equipment, installation, required breaks, scheduling of equipment operators, smoking, and so on. The new information management tools to be examined in this book hold the promise of giving designers and building managers the ability to respond to legal requirements, to contractual requirements, to employees' work-related needs, and to individual preferences.

Change of Occupancy: Although change is a given in our culture, we still build buildings much as we did centuries ago: very slowly, out of small parts, and with relatively little standardization or systems organization. Considerable systems thinking has gone into the process of construction (critical path methods, for example), but less has gone into the product. Perhaps the most striking characteristic of the built environment is its lack of responsive change. Buildings have a normal life cycle of at least forty years, yet the occupancies for which they are designed often change 100 percent by the time they are first occupied.

As was noted previously, the International Facility Management Association (IFMA) describes the portion of an organization that moves or requires facility changes each year in terms of a churn rate. IFMA research notes that a 30 percent churn rate is typical for a majority of organizations. This means that, on average, the entire occupancy of a building changes ten times over its (40-year) design life. Designers and space managers address these problems, but in a reactionary rather than anticipatory way that relies on systems furniture, open-office landscaping,

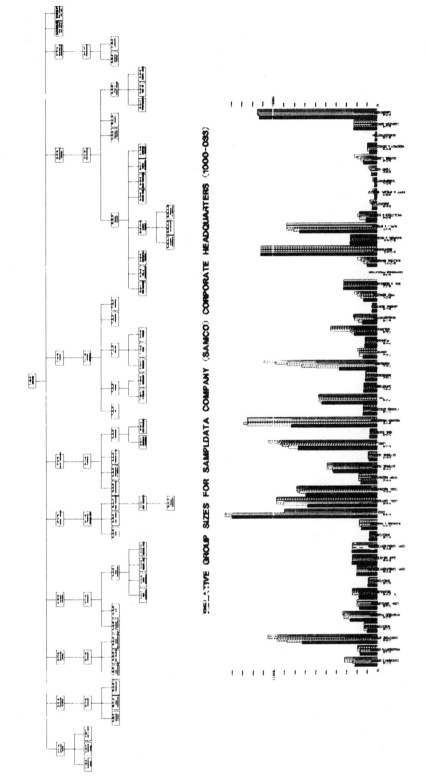

Figure 1-8. Typical organization chart. (Courtesy of HOK Computer Service Corp.)

A Facility Management Overview

13

and so on. These are essentially ways of attempting to retrofit flexibility into a long-lived and highly inflexible physical object: a building.

Information management tools give us the ability to anticipate change more effectively. Moreover, when reaction is the only course available, they give us the ability to react faster.

The speed at which change occurs has been a problem for facility designers and managers, and both the amount of change and the rate at which it occurs are likely to increase in the future. Industry and institutions are learning to change rapidly in order to survive and grow in the variable climate of governmental, social, and market forces.

Change to some extent reflects disorganization, or an inability to successfully anticipate the future. Disorganized change is unplanned. The corporations and institutions that inhabit buildings always exhibit some degree of disorganized (reactive rather than anticipative) behavior because we cannot hope to preordain all change or even to understand precisely why it occurs.

Incomplete Information: Facility managers seldom have complete information with which to plan. This is not due to their being unable to identify and gather information (as challenging a task as that is procedurally in any large organization); more fundamentally, top management intentionally withholds certain information from the facility manager.

Corporate, governmental, and institutional top managements are privy to information about the future, some of which is confidential. An example would be the pending sale or acquisition of a division or organizational unit that is still under negotiation and must remain secret.

Facility managers (or more precisely, space programmers) develop an architectural program for a new building, major move, or remodeling operation. Sometimes called the *space program* (or simply the *program*), the architectural program is a document specifying the needs of the occupancy: space and equipment, timing, lo-

cation, supporting services, and so on. The space programmer compiles and develops this needs document.

When interviewed by the space programmer, a corporate executive who knows that the company is about to double its size via acquisition will not tell the programmer, in order to maintain secrecy. Consequently, the building is designed for a particular organizational configuration that will never occupy it. The cost of an improperly designed facility or one that must be totally redesigned is in this case acceptable relative to the larger strategic costs avoided.

Management over Time: Time is significant in management plans and tasks. Saving time in facility management tasks is important, of course, but working with time effectively is even more important. It is insufficient, for example, to produce a locational siting plan or building vertical stacking plan optimized for a single plan date in the future. Tools are needed for producing plans that move from existing occupancies to future ones along the least-cost path of number (and cost) of moves. In response to this need, a few tools are beginning to emerge that recognize time and that help facilities managers program, plan, and manage over and through it.

Politics: Most organizations are to some extent political; many are heavily or even primarily political. It is appropriate in facility management to enlarge the traditional definition of politics to include personal and organizational goals (especially strategic ones) that cannot be straightforwardly and/or publicly stated. A more traditional definition of *organizational politics* might include individual and group conflict, "empire building," and competition for resources and status such as space.

Tools that help the designer/manager separate objective needs from more subjective or political ones are useful and available. The most straightforward are adjacency needs programming tools that filter requests—weighting mutual (two-way) interactions more heavily than individual (one-way) requests, and weighting "top–down" (manager-to-subordinate) interactions more heavily than "bottom–up" interactions. Even though the decision to satisfy political requests—or even to prioritize certain groups' subjective needs above others—is often validly made, tools are available to make such decisions explicitly rather than unknowingly.

The challenges, though many, are far from insurmountable. Subsequent chapters of this book examine how today's facility management tools and techniques can meet and overcome these challenges.

TWO

FACILITY MANAGEMENT TOOLS

A facility management system may be relatively simple or complex, manual or computer-assisted. It may be usefully depicted in terms of two axes (fig. 2-1): the horizontal axis indicates the level of complexity and size of the system; the vertical axis indicates the level of automation (the amount of integration/centralization of the databases, the number of simultaneous users/departments, and so on).

TRADITIONAL TOOLS

The traditional tools of facility managers have been the file cabinet and file folders of leases,

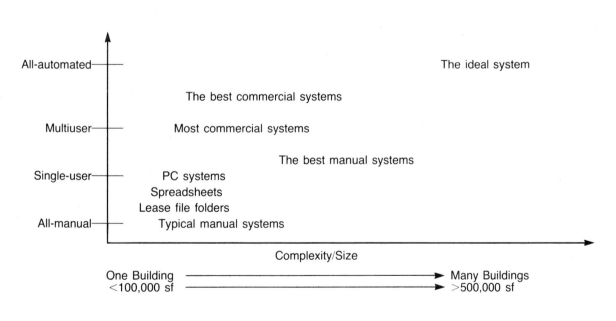

Figure 2-1. Taxonomy of facility management systems.

followed closely by architectural floor-plan drawings, file cards, and manufacturers' catalogs. For many individuals working in the field today, these remain the basic tools of a facility management system that consists of manilla folders bulging with equipment lists, personnel allocations, space projections, lease expiration dates, estimated square footages of existing facilities, and related forms, documents, and drawings of many kinds. It is an economical and effective system, limited only as to the size of facilities that it can be used to manage practically, given constraints on human speed and accuracy.

AUTOMATION

Early Information-management Tools

Since the late 1960s, use of data-processing technology to handle space, furniture, fixtures, equipment, and other fixed-asset inventories, as well as architectural programming data, has become increasingly common. Operations research methods have been applied in some contexts to space planning and management problems. But lack of adequately detailed current data and generally low levels of understanding by space planners and managers of these techniques have made their use sporadic and unevenly successful. Examples of facility management reports from such database management systems (DBMS) are shown in figures 2-2 and 2-3.

Computer-Aided Design and Drafting (CADD) Systems

Over the last several years, a second generation of design and drafting tools has emerged, as drafting system vendors have increasingly targeted space planning and management as an area of software applications and sales. Generally, their approach has been to extend a standard computer-aided drafting system by providing capabilities for automated takeoff of

BANKING - INTERNATIONAL

01 : BANKING - INTERNATIONAL

DL-1 SUMMARY LISTI

DEPT.	NAME	PRES	1986M	1989
01 A	INTERNATIONAL BANKING			
01 A01	ADMINISTRATION	2653.	3055.	3055.
01 A02	TREASURY	1595.	2035.	2567.
01 A03	OPERATIONS	2498.	2897.	3161.
01 A04	CONTROLLERS	683.	1005.	1005.
01 A05	PARTNERSHIPS	1284.	1939.	2300.
01 A06	TRADE FINANCE	818.	1418.	1740.
	GROUP TOTAL	9531.	12349.	13828.
01 B	IND RESOURCES			
01 B01	MANAGEMENT & TRADING	741.	1711.	1711.
	GROUP TOTAL	741.	1711.	1711.
01 C	ACC'T. SYSTEMS & OPERATIONS			
01 C01	SYSTEMS & OPERATIONS	5498.	8292.	9231.
	GROUP TOTAL	5498.	8292.	9231.

Figure 2-2. Facility management report. (Courtesy of Skidmore, Owings & Merrill)

areas and counts of furniture and equipment from floor-plan drawings, which allow the system to generate databases of space and to take inventories of furniture and equipment.

The most common additions to drafting systems (for purposes of turning them into facility management systems) include the following:

- Specialized menus containing facility management terminology and/or commands
- Furniture and/or equipment libraries containing drawings and costs of elements
- Furniture and/or equipment counting functions
- Area takeoffs (the ability to graphically "point out" and calculate an area)
- Stacking plan (a vertical section through a multistory building, showing occupancies) capability
- Block plan (schematic floor plan) capability

Computer-aided design (CAD), computer-aided design and manufacturing (CAD/CAM), Computer-integrated manufacturing (CIM), and computer-aided design and drafting (CADD) systems are attempts to integrate many functions into one hardware-software package. Often

DATE   7/14/86   PAGE 3

BUILDING CODE   RT1   BUILDING NAME   REDBUD TRAIL BUILDING #1

FLOOR	GROSS AREA	ZONE CODE	ZONE AREA	SPACE CLASSIFICATION	PLANNING UNIT CODE	PLANNING UNIT NAME	AREA OCCUPIED
2	81200.702	A1	22236.829	GENERAL OFFICE	P100030	NEWHOUSE CONSULTING	16034.225
					P100174	EFT GROUP	4002.604
		A2	18319.206	GENERAL OFFICE	P100209	MILLIKIN	6251.211
					P100633	PI INFORMATION SYSTEMS	9760.053
		B1	22366.567	WAREHOUSE	R100439	WESELEY PRINT SHOP	4009.122
					R100440	STOCK STORAGE	8002.452
					R100441	DISTRIBUTION PARTNERSHIP	3015.348
					R100893	VACATED 7/86	4987.893
		B2	18278.100	UNIMPROVED	T20004	WESELEY CORPORATION	15978.100
FLOOR TOTAL	81200.702		81200.702				72041.008
BUILDING TOTAL	197345.228		197345.228				156856.022

Figure 2-3. Facility management report. (Courtesy of Intergraph Corp.)

these are presented and sold as turnkey packages from a single vendor. (The term *turnkey* erroneously implies that the user simply "turns the key" and the package works.) Others are assembled by the user who combines hardware from one or more vendors with software from one or more others. (Figures 2-4 and 2-5, as well as figures C-1 through C-7 of the color insert, illustrate CADD systems from various vendors and show some of their outputs that are applicable to facility management.

The most common feature of a CADD system is what might be called a picture processor (by analogy to a word processor). CADD systems attempt to give their users much the same power and flexibility to edit and recombine pictures that a good word processor provides for text editing. Good CADD systems attempt to maintain not just pictorial representations but a geometric model. In this case, the user edits pictorial representations, but the system maintains only one consistent geometric representation.

The goal of large systems is shared access by many users. Charles Eastman, in his book *Spatial Synthesis in Computer-Aided Building Design*, has characterized the goal of such systems as "design information." Instead of copies of a design, he says,

All members of the design team may access common data through a timeshared computer database. Each can access the same or different information, organized and displayed in the manner most useful to that design specialty. Each may have one or more alternatives that are being developed for a part of the design that is a modification or extension to the stored data. When a designer thinks that an alternative is worth "adding to the permanent design," he or she checks with the other designers on the computer system and they can either visually review it or run a program that compares it with their own working alternatives.

Some CADD systems can establish relationships between pictures and text. For example,

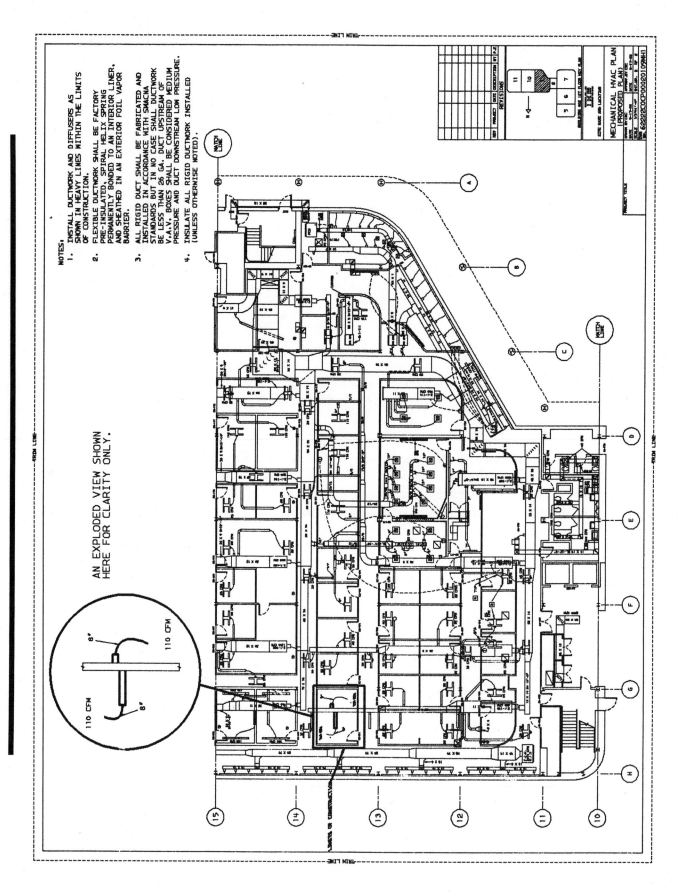

Figure 2-4. CADD output. (Courtesy of IBM Corp.)

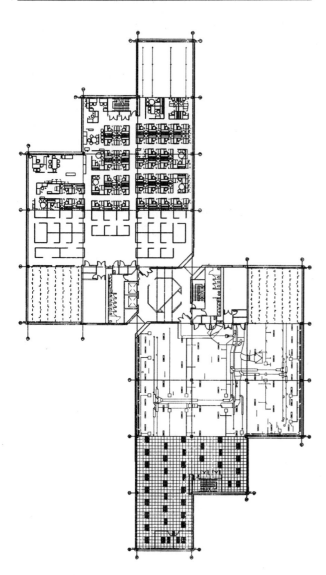

Figure 2-5. CADD output. (Courtesy of McDonnell Douglas Corp.)

totals, takeoffs, and so on) to be exported in machine-readable form to a word processor, data base management system (DBMS), or other external applications program. This capability could be used simply for incorporation into text documents such as bids and reports, or it could be used as a data collection mechanism for a comprehensive integrated facility management system.

Applications Programs

The next stage in the evolution of computer-based facility management systems was the appearance of applications programs designed specifically for facility managers. Organizations today are offered a wide range of computer applications programs to support their facility management efforts. Some applications support the design and planning processes; other support ongoing operations. These ad hoc or stand-alone packages are designed to solve one particular type of facility management problem and operate "standing alone," apart from coordination with other programs. Stand-alone packages now cover the entire range of the design process, from initial collection of occupant needs data to feasibility analysis to construction scheduling and supervision to operations management and maintenance. General software tools can be used to develop custom applications such as electronic spreadsheets, database management systems, application generators, and text processing.

Discrete stand-alone facility management programs have certain disadvantages, foremost among which is lack of coordination (sharing of data). By definition stand-alone programs were never designed or meant to coordinate (data or otherwise) with other programs. They were designed to solve one particular problem. This makes cost-effectiveness much more difficult to achieve, since the cost of building the information base must be recovered through the benefits obtained from the (only) application.

the name of the manufacturer of a chair can be stored in a CADD system as a string of text, and related to the picture of that chair. The chair's manufacturer can thereafter be recalled (with the picture of it) if and when desired. While CADD systems generally do not possess extraordinarily sophisticated word-processing capabilities, a few allow generated text (counts,

COMPREHENSIVE INTEGRATED FACILITY MANAGEMENT SYSTEMS

A *comprehensive* facility management tool is one that addresses or incorporates all (or at least a large majority) of the disparate facility management activities and functions identified in chapter 1 of this book.

An *integrated* facility management system is one in which information from any one of these many activities (facility management applications) is available for use in all others, without requiring user programming or reentry of data. Thus, in the view of the user, an integrated facility management system has only one database. (The implementation may utilize separate physical or logical files, as long as their transparency to the user of the system is maintained.)

Accuracy/Validity

Good facility management systems (like good computer-based information systems for any other application) have both form and substance. The form, format, quality, and comprehensibility of (often graphic) output is important. This form is critical to the ability of (often time-pressured) upper management personnel to assimilate the message and act.

Still more important is content. The substance (accuracy, correctness, and validity) of the information displayed (especially if it is displayed in "easy-to-understand" decision-format graphics) is critical. Too often, computer output is assumed to be correct, without the questioning that would routinely be applied to manually aggregated and typewritten information. The algorithms underlying and written into computer programs are sometimes crude (because such algorithms were easier to program), inaccurate, or sloppy. Occasionally, they are just plain wrong. Too often, managers fall victim to the old data-processing maxim, GIGO (garbage in, garbage out—a way of saying that,

even when the computer program is correct, if the input data are poor the result will be poor).

It is easy to confuse form with substance. Three-dimensional color views of proposed space are beautiful, but they rarely provide the bottom-line impact of a warning to exercise an upcoming lease option. The important things to keep in mind are the two goal benefits discussed earlier: (1) to reduce cost, and (2) to improve facility productivity by better fitting the building to its occupants. There is no harm in beautiful graphics (and sometimes there is benefit), but content is most important. The objective is decision support that achieves the above two goals.

Often the medium becomes the message. This is especially easy and all too common with computer graphics. Decisions that save money and improve the match of facility needs with facility resources seldom result from high-resolution three-dimensional color renderings of interior space. More often, good decisions result from a few facts presented in a clear format. This can be a pie or bar chart, or a few numbers. Simplicity is often the key. Mies Van der Rohe's often-quoted comment about architectural style, "Less is more," applies to presentation formats, too.

The following lessons can be deduced from the preceding points:

- Good procedures are good, with or without a computer.
- Question, investigate, and prove to your satisfaction the correctness of computer programs.
- Question and investigate the practical value and usability of flashy graphics. Seek formats (graphic and otherwise) that support informed decision-making.

Characteristics

Many organizations today are just beginning to implement and aggressively use comprehensive integrated computer systems to perform facility management functions. The words, *computer systems* rather than *computer software* are used in the preceding sentence because much more than computer code is required to ensure the facility manager's success as a computer user. A facility management system must include the following elements:

- *Procedures,* usable with or without a computer
- Initial and ongoing *education* and *training*
- Computer *software*
- Exquisitely thorough procedural and software *documentation*
- Ongoing *support* vehicles (regular updates, "bug" fix mechanisms, user groups, telephone, and so on)

Some of the first attempts at comprehensive integrated facility management systems were developed in-house over a period of years, just as early CADD systems were. Some of these early systems were developed by architectural or interiors firms, others by corporations or institutions. Commercial products embodying comprehensiveness and integration are now beginning to emerge as well. Such a facility management system (as opposed to a series of separate software packages) has the following basic characteristics:

- *Comprehensiveness:* All application areas of facility management are addressed.
- *Integration of data:* A single information source (database) is available to each of the application areas and (within desired security restrictions controlled by the user organization) to all users. Information is entered only once, so there is no redundancy of data.
- *Modularity:* The various application areas are separate and can be acquired serially and used as discrete programs, even though they share a common database.
- *Flexible procedures:* Thorough procedures (illustrating the tasks, with or without automation) are incorporated. Specific, detailed direction for such tasks as programming, planning, design, construction, rearrangement, leasing, maintenance, and procurement are given for decision making.
- *Decision support:* Output information is formatted

for anticipated decision making (as opposed to being formatted for implementing decisions).

- *Ease of use:* The wide range of potential users of a facility management system makes ease of use critical. Facility management staffs are generally new to automation. Management is an infrequent, often poorly trained, but very important user. Techniques such as screen forms to walk the user through each step in the process, online multilevel HELP keys, and so on, should be available at every point, with clear examples and explanations. Experienced users, however, should be able to bypass unneeded prompts and explanations.

Various designs for integrated facility management software systems are schematically illustrated in figures 2-6 through 2-11.

Components of a Facility Management System

This section describes key individual components of a comprehensive facility management system, by function. Some organizations will utilize only a few of these components; most (over time) will find applications for all of them.

Inventory Management: This involves developing a comprehensive database of space, personnel, and equipment, as well as of activities and organizations. It should include occupancy and alternative-use space characteristics, and it should be organizable along either physical (city, site, building, floor, room, and so on) or organization-chart lines. It may cover tracking of associated unit costs, billing, lease information, or other such data.

Requirements Programming: This involves compiling information for and making projections of future facilities needs in all of the areas discussed in the inventory management component. The basis for projection should include standard and user-defined *defaults* (values to be used by the system if none is specified explicitly by the user), as well as the ability to use historical, economic, demographic, or pro-

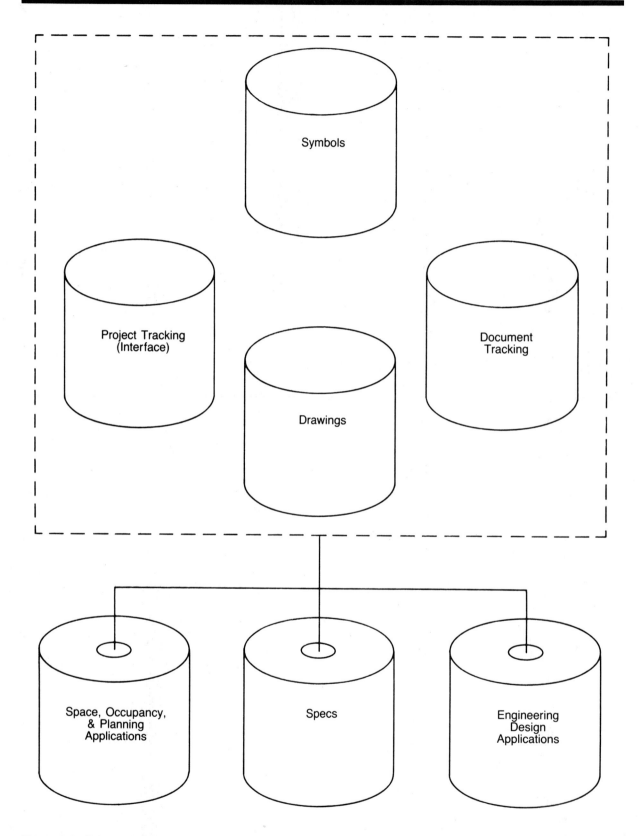

Figure 2-6. Schematic design for a facility management system. (Courtesy of IBM Corp.)

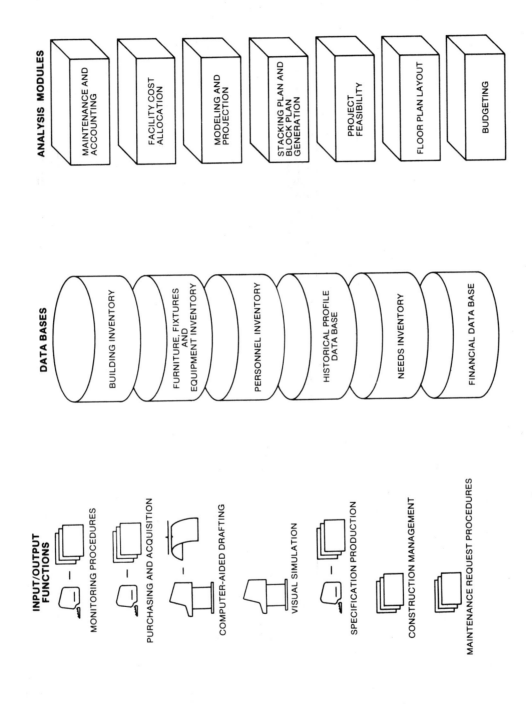

ANALYSIS MODULES

- MAINTENANCE AND ACCOUNTING
- FACILITY COST ALLOCATION
- MODELING AND PROJECTION
- STACKING PLAN AND BLOCK PLAN GENERATION
- PROJECT FEASIBILITY
- FLOOR PLAN LAYOUT
- BUDGETING

DATA BASES

- BUILDING INVENTORY
- FURNITURE, FIXTURES AND EQUIPMENT INVENTORY
- PERSONNEL INVENTORY
- HISTORICAL PROFILE DATA BASE
- NEEDS INVENTORY
- FINANCIAL DATA BASE

INPUT/OUTPUT FUNCTIONS

- MONITORING PROCEDURES
- PURCHASING AND ACQUISITION
- COMPUTER-AIDED DRAFTING
- VISUAL SIMULATION
- SPECIFICATION PRODUCTION
- CONSTRUCTION MANAGEMENT
- MAINTENANCE REQUEST PROCEDURES

Figure 2-7. Schematic design for a facility management system. (Courtesy of Computer-Aided Design Group)

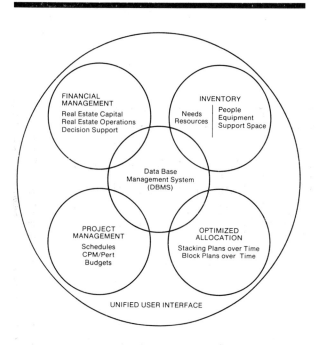

FINANCIAL
MANAGEMENT

Real Estate Capital
Real Estate Operations
Decision Support

INVENTORY

Needs
Resources

People
Equipment
Support Space

Data Base
Management System
(DBMS)

PROJECT
MANAGEMENT

Schedules
CPM/Pert
Budgets

OPTIMIZED
ALLOCATION

Stacking Plans over Time
Block Plans over Time

UNIFIED USER INTERFACE

Figure 2-8. Schematic design for a facility management system. (Courtesy of Computer-Aided Design Group)

duction data. It should be possible to attach attributes (generalized characteristics such as number of parking spaces required, environmental effects, electrical needs, and special financial, legal, or personnel constraints) to any of the items projected (space, personnel, or equipment needs).

Master Planning: This involves preparing phased strategic plans that satisfy long-range space needs and estimating associated costs. The primary purpose of master planning is to anticipate and respond to *business changes:* to analyze options and ask high-level "what if?" questions based on cost, time, and availability of resources. Master planning may be thought of as a matrix in which one axis is needs (activities, departments, and so on) and the other axis is resources (buildings, equipment, and so on); each cell of the matrix contains information about what is and what is not possible, together

with the associated costs and the time required. There can be many such matrices (one for each time period or alternative scenario under consideration). Included in this component are anticipating and documenting required intermediate moves, with resulting acquisitions, leaseholdings, required construction, and so on.

Location and Layout Planning: This involves preparing optimized city and site location decisions, multistory building stacking plans, and block (schematic) floor plans. For this purpose, the inventory and requirements database information is enlisted: area (required and available), adjacency requirements, fixed costs (such as for moving or required construction), preassignments, and preferences. Phasing the optimized result to achieve a least-cost move path is done in this stage.

Drafting: This involves interfacing with CADD systems, which are the tool of choice for implementing decisions reached by means of an integrated facility management system. CADD systems work well in an integrated environment because of the possibility of transferring much of the required information electronically from one application to another. An example is the automatic availability of architectural program information (for instance, "How many file cabinets need to be layed out for this department?") during CADD drafting for space planning and layout. With electronic availability of information come the attendant benefits of time and cost savings, as well as reduction of errors. Information related to space occupancy, areas, names, assignments, and so on, are passed back and forth between the facility management database and the construction documents produced by the CADD system. Although a fully integrated system ought, perhaps, to incorporate a CADD system closely coupled with the rest of the database, it may not be immediately practical to do this because so many competent industry-standard drafting systems are in use already. A checking feature would ensure that the plans accurately represent solutions, as de-

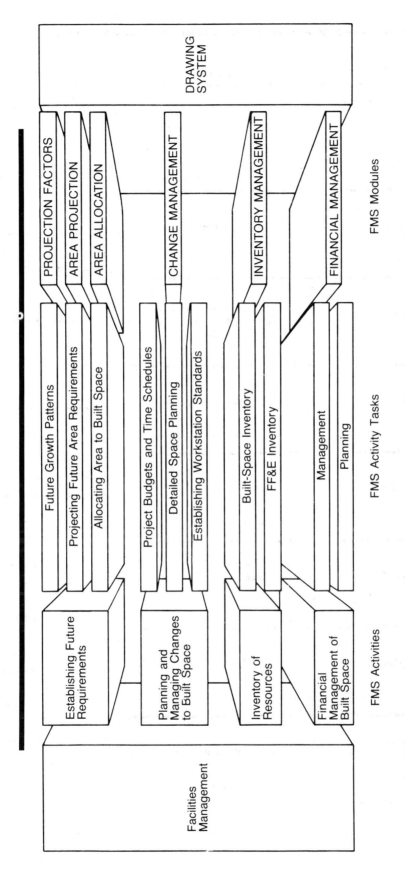

Figure 2-9. Schematic design for a facility management system. (Courtesy of Computer-Aided Design Group)

Input/Output Function	Databases	Specific Applications
Monitoring Procedures		
Purchasing and Acquisition	Building Inventory	
Floorplan Layout	Furniture, Fixtures, and Equipment Inventory	Maintenance and Accounting
Computer-aided Drafting	Personnel Inventory	Facility Cost Allocation
Visual Simulation	Historical Profile	Modeling and Projection
Specification Production	Needs Inventory	Stacking Plan and Block Planning Generation
Construction Management	Financial	Project Feasibility
Maintenance Request Procedures	Alternate Scenarios	Budgeting
Report Writer		

Figure 2-10. Schematic design for a facility management system. (Courtesy of Computer-Aided Design Group and Richard Dilday)

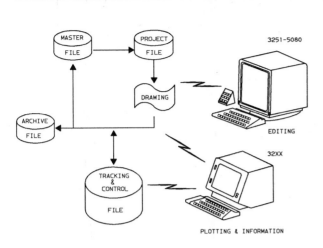

MASTER FILE → PROJECT FILE

3251-5080

DRAWING

EDITING

ARCHIVE FILE

32XX

TRACKING & CONTROL FILE

PLOTTING & INFORMATION

Figure 2-11. Schematic design for a facility management system. (Courtesy of IBM Corp.)

fined in the requirements program. Conversely, a CADD interface ensures that the information in the plans can be used for additional planning or analysis.

Cost Accounting: This involves aggregating and allocating operating cost and capital cost by space, person activity, and so on, as well as preparing and apportioning budgets to the various billing numbers, projects, and cost centers in the organization.

Real-Estate Strategy: This involves compiling real-estate information (construction costs for building types, detailed lease information, and projected acquisition costs for alternative scenarios) and formulating recommendations based on that information (lease, build, buy, move, and the like).

Move Coordination: This involves minimizing disruption through careful planning, staging, and coordination of moves and move sequences—including furniture and equipment tagging and staging.

Project Administration and Implementation: This involves coordinating all tasks in facility

management, with the primary aim of managing resources (personnel, time, and money). Included are project budgeting, resource management, and project scheduling, as well as schedule coordination, construction cost allocation, and job tracking.

Purchasing Coordination: This involves generating purchase orders that have been cross-checked with inventory, allocating costs, and coordinating costs with fixed assets inventory and depreciation.

Maintenance Planning: This involves producing maintenance schedules for equipment within the facilities and for the physical facilities themselves, keeping records of unplanned maintenance and consequent replanning of preventive maintenance, and allocating costs incurred in maintenance.

Site Management: This involves tracking and managing utilities and infrastructures, as well as managing a database of their dependency relationships.

System Coordination: This involves integrating all information required or generated by all of the preceding functions, including backup, journaling, audit trails, error and disaster recovery, and general support functions (database administration, presentation graphics, and so on).

FUTURE STEPS

The evolution of facility management software will certainly not end with the sophisticated integrated facility management packages that are beginning to appear on today's market. Some beneficial developments we may hope to see are discussed in this section.

Standard Interfaces: Many organization-wide databases need to transfer information to (and receive information from) a facility management system. This may not occur until anticipated shakeouts and standardization occur in the computer industry. At that point, facility management systems should be able to tap into databases such as personnel, fixed assets, and strategic planning, without any additional programming, consulting, or interfacing. At least two standard interfaces (dealing with CADD systems) have already begun to emerge. The IGES (Initial Graphic Exchange Standard) is, as the name suggests, a standard interface for exchanging graphics information between CADD systems. IGES is utilized, at least in primitive form, by most CADD companies. The Facility Drafting Coordinator, from the Computer-Aided Design Group, is a standard interface for the information and interaction shared between CADD systems and facility decision-support systems. Many major CADD firms are exploring ways of utilizing this standard.

Increased Decision Support: The use of *truth tables* (matrices of situations and decisions whose validity or invalidity may be inferred in each circumstance) and similar devices will probably be the first step toward the application of artificial intelligence.

Expert Systems: This valid (but unfortunately much-exaggerated) phenomenon will affect facility management applications just as it will all other computer applications. Facilities managers will use expert systems to suggest moves, financial strategies, and (ultimately) solutions to space-planning, utilization, and even aesthetic problems.

Expanding Pallette of Applications: Already facility management shows signs of expanding into the following applications areas:

• Mapping
• On-line energy conservation
• Automated maintenance
• Interactive building monitoring

More applications will surely follow.

THREE

COSTS AND BENEFITS

Before an organization can decide intelligently to implement facility management systems or to expand/improve existing procedures, a cost/benefit analysis and an associated return on investment (ROI) analysis are appropriate.

Additionally, when an organization decides to purchase a computerized facility management system, it is faced with a wide range of choices. Almost always, even when the list of applications has been narrowed, too many applications remain to enable the organization to implement them all in a single simultaneous effort. Budgetary constraints are inevitably an issue. Even if money were no object, clear human limits restrict how much newness (procedures and tools) can be assimilated at one time.

To plan intelligently for the use of computer tools—as well as to place them in a priority order—the organization must use a rational process. An important part of this process is *cost/benefit analysis,* whose goal is to identify the applications that provide the greatest return for the investment required. The idea is to look at the ratio of all of the benefits (return) to all of the costs (investment) for a given application over an equivalent time period (usually a year). Using this approach produces a list of applications that can be ordered by the value of the

benefit-to-cost ratio of each. Figure 3-1 shows a sample list of potential facility management applications for one organization, ordered by the simple ratio of total benefits to total costs. Note that where the benefit-to-cost ratio goes below 1.0, the return on investment is negative; that is, a benefit-to-cost ratio of 1.0 (not 0.0) is the break-even point. Below 1.0 the net economic result of implementing the application is negative.

A "stop implementation" line generally should be drawn at the first of the following three points to arise in any ordered listing of applications options:

- The point at which total costs exceed budget
- The point at which the organization feels it cannot implement more, as a result of organizational issues such as training and personnel time
- The point at which the return on investment no longer is positive

Note that in figure 3-1, CADD did not show a positive return on investment, though it is an established and generally cost-effective technology. In this particular example, however, the analysis was done specifically for facility management tasks, for which CADD was not cost-effective. CADD would have been cost-effective if the organization did most of its im-

APPLICATION NAME	TOTAL ANNUALIZED COSTS	TOTAL ANNUALIZED BENEFIT	NET ANNUAL GAIN OR ⟨LOSS⟩	BENEFIT-TO-COST RATIO
Space Inventory	$ 28,000	$197,000	$169,000	7.04
Lease Tracking	$ 19,000	$132,000	$113,000	6.95
Cable Management	$ 19,000	$ 96,000	$ 77,000	5.05
Equipment Inventory	$ 39,000	$113,000	$ 74,000	2.90
Strategic Planning	$ 21,000	$ 45,000	$ 24,000	2.14
Schematic Planning	$ 16,000	$ 18,000	$ 2,000	1.12
CADD Drawings	$132,000	$130,000	⟨$ 2,000⟩	0.98
Budgeting	$ 18,000	$ 17,000	⟨$ 1,000⟩	0.94
Project Management	$ 43,000	$ 40,000	⟨$ 3,000⟩	0.93
Purchasing	$ 18,000	$ 15,000	⟨$ 3,000⟩	0.83
3D Simulation	$185,000	$ 40,000	⟨$145,000⟩	0.22

Figure 3-1. Applications by benefit/cost ratio.

plementation in-house instead of using architectural firms for its major building projects.

ASSESSING COSTS

The *life-cycle cost* of a system is the total of all costs it incurs over its lifetime. Such costs are paid out over time (over the life cycle of the system). Simply totaling these costs, however, ignores the time value (or *opportunity cost*) of money. The most appropriate way to look at life-cycle costs is to take the present value of the stream of payments, discounted by the expected opportunity cost of money (interest rate or *discount rate*) over time. The resulting figure is called the *present-value life-cycle cost* (*PVLCC*). A simple example is shown in figure 3-2, in which the discount factor column represents the value (or cost) today of money received (or paid) in the future. It begins at 1.00 (1 dollar multiplied by a discount factor of 1.00) and is reduced by the discount rate, which in this example is 8 percent annually. The discount factor in each subsequent year is calculated by dividing the previous year's discount rate by

Parameters
15-year term
Costs: $100,000 per year
Discount rate: 8 percent per year

YEAR	DISCOUNT FACTOR	PAYMENT	PRESENT COST
1	1.00	$ 100,000	$100,000
2	0.93	$ 100,000	$ 92,593
3	0.86	$ 100,000	$ 85,734
4	0.79	$ 100,000	$ 79,383
5	0.74	$ 100,000	$ 73,503
6	0.68	$ 100,000	$ 68,058
7	0.63	$ 100,000	$ 63,017
8	0.58	$ 100,000	$ 58,349
9	0.54	$ 100,000	$ 54,027
10	0.50	$ 100,000	$ 50,025
11	0.46	$ 100,000	$ 46,319
12	0.43	$ 100,000	$ 42,888
13	0.40	$ 100,000	$ 39,711
14	0.37	$ 100,000	$ 36,770
15	0.34	$ 100,000	$ 34,046
TOTAL		$1,500,000	$924,424

Present value of 15 years' $100,000 payments is $924,424.

Figure 3-2. Simple illustration of present value.

1.08 (for the 8 percent). This yields a discount factor of 0.93 in year two (1.00 divided by 1.08), which is another way of saying that at 8 percent interest, a dollar next year is only worth 93 cents today. In year three the discount factor is 0.86 (0.93 divided by 1.08) and so on for the following years.

The significant measurable cost components of computer use in facility management (and in most other computer applications, as well) may be separated into hardware-related costs (the smallest amount of money over a system's life cycle), software-related costs (larger), and information-related costs (the largest).

Hardware Costs

Hardware costs are the most immediately obvious costs and are likely to loom large in the minds of the organization's top management who are about to make a major outlay for acquisition of a system. Nonetheless, hardware costs represent only the tip of the iceberg of total costs. Hardware costs take the form of rental, leasing, or purchase and finance costs on in-house equipment, or they may represent a component of timesharing or service bureau fees.

Software Costs

Software costs, another major cost component, usually greatly outweigh hardware costs in the long run. Software costs can include the following:

• *Purchase* costs for acquiring items from software vendors
• *Royalties* and *surcharges* for use of software, charged either at a fixed (often monthly) rate or according to the amount of usage
• *Research and development* costs, either in the form of in-house staff costs or as payments to consultants, where custom software development (or major adaptation) is undertaken
• *Maintenance* costs in the form of maintenance contract payments to the vendor/developer or

similar in-house staff costs
Introduction costs in the form of staff time for the preparation, installation, verification, and maintenance of practically all new software acquisitions

It is easy to underestimate the true costs of in-house software development or of major in-house modification of existing software. Software production managers tend to assume that the only major cost involved in this process is that of creating the program code; all but the most experienced programmers underestimate development time and cost. A sound piece of production software, however, requires very careful system specification and design, involving extensive participation by the intended users. Additional requirements include ongoing maintenance, training, support, and enhancement efforts, all of which must be carried out by in-house staff or outside parties. These additional staff costs—particularly if they are unanticipated—can be crippling to the organization. Furthermore, software development and maintenance efforts require considerable management attention. Experience has shown that, over the life cycle of a piece of production software, the average expenditure allocations break down as follows:

SPECIFICATION AND DESIGN	33–40%
PRODUCTION (CODING) AND TESTING	20–33%
SUPPORT AND MAINTENANCE	33–40%

Operations Costs

Operations costs are mostly associated with people: those who attend to the computer equipment; those who find, enter, edit, and verify data; those who offer consultation to users; and those who provide support. The cost of gathering, loading, and correcting information in computer operations is great.

Data Entry Costs: The largest single component of the life-cycle cost of any computer application is the cost of data entry: entering data into the computer correctly and in a form that is machine-readable and relatively error-free; identifying the correct sources of the information; filling out forms; performing key or graphic data entry; and so on.

It is not unusual for data entry costs to exceed all other costs combined! This strongly implies that the organization implementing automation of any kind (especially supposedly small or inexpensive systems) should focus not on the cost of the hardware or even of the software, but on two operational areas:

- The quality of the procedures (both internally developed and vendor-supplied) for entering, verifying, correcting, and maintaining information over time
- The power, ease-of-use, and robustness of the automated and manual systems that are designed to support data acquisition and maintenance

Attempting to save money on software or hardware (supposing that it affects the cost or ease of data entry and ongoing maintenance in any way) is absolutely an example of false economy and is usually a very costly mistake.

Education and User Liaison Costs: Education and user liaison costs comprise the costs of training staff within the organization to use systems, procedures, equipment, and software; to prepare data; to interpret output; to act on results; and so on. These costs are substantial and usually represent a fairly risky investment. Staff trained in good facility management procedures and skills *or* in computer skills are in great demand, and recovering the training cost of a staff member who leaves is impossible. This means that good-quality facility management staff must receive good-quality compensation, if they are to be retained. The impact of CADD on the design professions provides a good model: good architects who have training in CADD skills are at present highly sought after. While it is to be expected that computer skills and experience will become increasingly common in the design disciplines, high-quality computer applications for facility management

are several years behind those for design. For this reason, good facility managers with computer experience will be competitively sought after for many years to come. Top management must recognize this and plan appropriate compensation programs for such managers or suffer high turnover and resultingly high training and education costs.

Management Costs: Management costs accrue from time spent in planning, implementing, and monitoring facility management activities. These costs, too, can be considerable, since a successful facility management operation develops into a major area of capital investment and planning, with substantial staffing requirements.

Overhead Costs: Computer-related overhead includes costs of space, air conditioning, power, and supplies. Space costs are not normally a major concern, since modern computer equipment is small, but power and air-conditioning costs can be surprisingly high for larger systems. Supply costs for items such as plotter and printer paper, tapes, disks, and ribbons have a tendency to get out of control, in the same way that photocopying costs do.

Intangible and Hidden Costs

Some costs are less obvious and are difficult to quantify, but they also should be considered.

Down-time Costs: All computer systems have the potential to fail ("go down") because of hardware and/or software problems. Where entire design, support, and decision-making processes depend heavily on a computer system, down time can be particularly costly, since work on a project, billing, report, lease analysis, or other activity may come to a halt when the system is down.

Costs of Correcting Bugs: The emergence of inconsistencies or errors ("bugs") in software or in supporting procedures can delay progress on a project and produce faulty work that must be redone.

Organizational Rigidities: Full potential benefits of a computer or procedural system may not be achievable, at least initially, because of interdepartmental political or procedural differences, recalcitrant personnel attitudes, and similar constraints on the possibilities for reorganization steps to take advantage of benefits.

Reduced Responsiveness to Workload Fluctuations: Introducing a computer system transforms an organization's facility management from a labor-intensive process into a much more capital-intensive one. In the traditional labor-intensive process, fluctuations in workload can be met (albeit with difficulty) by hiring and firing staff. But in the automated process, the amount and carrying costs of the hardware and software usually cannot be increased or reduced with comparable speed or effectiveness at times when the workload changes rapidly. Fortunately, well-organized procedural and computer systems support a significantly increased workload over traditional manual methods.

Another Way to Look at Costs

An alternative way to estimate costs is to compile them by category instead of by function. These can be expressed as percentage of purchase ("sticker") price of the hardware and of the software (capital). The categories are as follows.

Tangible Capital (Hardware and Software) Costs: The tangible capital costs can be further divided into three subcategories:

1. *Principal:* This subcategory covers the amortization of principal (the purchase price) over the expected useful life of the system's physical components. Conservative life expectancies would be three years for hardware and five years for software, assuming that both are properly maintained. This means that roughly 2 percent of the sticker

price per month should be budgeted for amortization of principal.

2. *Interest:* This subcategory covers the interest payments to a bank, leasing company, or similar lender, as well as the alternate-use-of-money cost for organizations fortunate enough to have available cash for a purchase. Interest rates vary widely, both with the times (inflation rate, cost of money, and so on) and with the organization (its ability to achieve a given rate of return on investments other than a facility management system). A good minimum rule of thumb for budgeting with respect to interest rates is 1 percent of the sticker price (12 percent per year simple interest). While not a predictor of future interest rates, of course, this rule of thumb is easy to use and, in most circumstances, accurate enough.

3. *Maintenance:* Computer hardware requires maintenance, just as automobiles and office copiers do. A maintenance contract (with the manufacturer, the distributor, or a third-party maintenance organization) is the simplest way to budget such expenses effectively. A good budget figure for maintenance costs is 1 percent of the sticker price per month. Software, too, requires updates, bug fixes, and user support. Established software vendors handle this in much the same way that hardware vendors do. The figure of 1 percent of the purchase price per month for software is similarly an adequate minimum budget figure.

The three budget items identified here as tangible capital costs total 4 percent of the purchase price (total of hardware and software) per month. By simple arithmetic,

$$4\% \times 12 \text{ months} = 48\% \text{ per year}$$

This means that simple "pride of ownership" (the privilege of owning—not using—any computer system) costs approximately one half of the initial purchase price of the system annually.

Operations and Personnel (Intangible Capital) Costs: These costs include the following items (in approximate descending order, by cost):

- Data input
- Data gathering
- Data preparation
- Education (ongoing)
- Education (initial/"learning curve")
- Conversion of existing systems
- Operators/operations
- System limitations/rigidities
- Down time
- Bugs
- Supplies usage/overhead

The experience of data processing consultants and users shows that operations and personnel costs equal or exceed tangible capital acquisition costs over a system's life cycle. This means that, if tangible capital acquisition costs amount to one half of the hardware and software purchase price each year, total life-cycle costs for the combined tangible and intangible elements are at least equal to the hardware and software sticker price, each year! The amount by which intangibles exceed tangibles varies with the type of application and how it is implemented. A data entry–intensive application, such as a manually updated furniture inventory, will have intangibles much greater than those of the same inventory application where the updates are gathered electronically from a computer database in another part of the organization, such as fixed assets.

The surprisingly high life-cycle cost of facility management systems (indeed of any computer application) is the bad news about automation. The good news is that the benefits, too, are unexpectedly large.

Operations and personnel costs do not vary proportionately with the system price tag. This is because costs such as training and education, establishment of procedures, and data entry are largely fixed, and therefore loom much larger as a percentage of the hardware and software purchase price to any organization attempting to cut costs by reducing the cost of the hardware and software system. Moreover, relatively large

hardware and relatively sophisticated software and procedures tend to reduce the amount of staff time spent on training, initial setup, establishing procedures, data entry, and so on.

Hardware and software are, therefore, the worst places to seek economies in a computer-based facilities management system. Procedures, information gathering, data entry, and similarly continual labor-intensive tasks are the best places.

Cost-estimation Forms

Figures 3-3 through 3-5 are examples of cost-estimation worksheets from the Computer-Aided

Microcomputer (3 workstations)
Costs for 500,000 Sq. Ft.

	One Time	Annual
Software Purchase	$ 5–20K	
Software Maintenance		$ 1–3K
Hardware Purchase	12–50K	
Hardware Maintenance		2–8K
Training	10–60K	5–10K
Data Entry (2 FTEs)	70K	10K
Consulting	30K	10K
	$127–230K	$27–41K

Mini or Mainframe (with drafting)
Costs for 500,000 Sq. Ft.

	One Time	Annual
Software Purchase	$160K	
Software Maintenance		$24K
Hardware Purchase ($50K to 80K) ($30K Increment for 500KSF)	150K	
Hardware Maintenance		22K
Training	80K	20K
Data Entry (4 FTEs)	140K	20K
Consulting	30K	10K
	$560K	$96K

Figure 3-3. Estimating costs. (Courtesy of Computer-Aided Design Group)

Software Costs

System Coordinator	$15K
Inventory	35K
Requirements Programmer	25K
Master Planner	15K
Location/Layout Planner	15K
Drafting Coordinator	5K
	$110K

Costs for 500,000 Sq. Ft.

	One Time	Annual
Software Purchase	$110K	
Software Maintenance		$18K
Hardware Purchase	75K	
($50K to 80K)		
($30K Increment for 500KSF)		
Hardware Maintenance		10K
Training	60K	10K
Data Entry (2 FTEs)	70K	10K
Consulting	20K	10K
	$335K	$58K

Annual Cost for Increased Square Footage

1M	$85
2M	$110
3M	$160

Figure 3-4. Total costs example. (Courtesy of Computer-Aided Design Group)

Design Group's seminar, "Benefits of Facility Management."

CATEGORIZING BENEFITS

The International Facility Management Association's membership in 1985 cited the benefits shown in figure 3-6 as resulting from their use of computers in facility management.

The benefits of automation in facility management can be classified according to the following categories:

- Increasing productivity
- Reducing response time on projects
- Enhancing design quality
- Reducing errors
- Enhancing management effectiveness
- Hidden benefits
- Smoothing out peaks and valleys
- Building the worth of the organization
- Taking advantage of system integration

These categories are discussed individually in the subsections that follow.

Increasing Productivity

The total time and effort spent on a typical building design or management function is distributed very unevenly across the stages of the

CUMULATIVE ($K)

SQ. FT. (K)	1 YEAR		2 YEARS*		3 YEARS*	
	COSTS	BENEFITS	COSTS	BENEFITS	COSTS	BENEFITS
500	335	282	393	564	451	846
1,000	435	564	520	1,128	605	1,692
2,000	635	1,128	745	2,256	855	3,384
3,000	835	1,692	995	3,384	1,155	5,076

* Does not reflect growth of company or third-year person.

Figure 3-5. Cumulative costs and benefits for different areas. (Courtesy of Computer-Aided Design Group)

Benefit	Percent Cited By
More data for decisions	67%
Increased accuracy	62%
Reduction of repetitive tasks	59%
Increased speed	52%
Expanded capabilities	40%
Reduction of personnel	6%
Other	6%
None	0%

Figure 3-6. Benefits of using computers. (Courtesy of International Facility Management Association)

design and management processes. Most effort is expended at the stages of information-gathering and data entry, where a great deal of detailed documentation is required. This implies that the productivity benefits of computer applications show up mostly in the early stages, where even a small percentage reduction in expenditure is substantial.

Productivity benefits can be achieved whenever labor is replaced by capital. Such benefits usually accrue in the form of reduced costs, as a result of such factors as the following:

- Reduction in the total professional, technical, and clerical hours of labor required
- Reduction in the amount of time required to complete the work
- Savings in the space required for personnel, desks, files, and so on
- Faster detection of problems, before they become costly
- Reduction of routine, clerical components in professional and technical staff jobs, and possible replacement of some higher-level jobs with lower-level jobs

Most of the benefits just named can be achieved with or without a computer. It is procedures that make the benefits available; com-

puters merely accelerate the delivery and accuracy of end products by executing most of the procedures faster (by several orders of magnitude) and with greater precision than manual methods can.

Reducing Response Time on Projects

A major potential benefit of effective computer use is the ability to complete a project in significantly less time. This may yield the following results:

- Increased capacity to accept tight-schedule projects that might otherwise not be taken on
- In the context of an inflationary economy, reduction in eventual cost
- Ability to make very rapid emergency revisions to designs—where required, for example, by unexpected conditions encountered during construction—and so to minimize losses resulting from such emergencies
- Capacity to support unexpected management requests with high-quality detailed information

Enhancing Design Quality

The management and design decisions having the greatest impact on costs and quality are usually those made early in the process. As the process continues, it focuses on refining an increasingly fixed concept and can have less and less impact on cost and quality. Most of the quality benefits of computer use are caused by software that applies to early stages of the design and decision-making processes.

Automated procedures serve to enhance design quality in a wide variety of ways. Two of the most visible are processing power (capacity for speed and detail), and consistency (predictability and correctness of repetitive operations). The speed and capacity for detail of procedures implemented using a computer allow designers to consider many more alternatives than might be explored manually. An example is in a space layout where the occupants' needs are displayed in great detail and organized in a manner which follows and supports the designer as he or she performs the design and space planning processes. The level of detail achievable and the speed with which detailed information is available to answer questions permit a design solution considerably more tailored to the needs of the future occupanyc than would be possible without automation.

The predictability and consistency of automated procedures greatly reduces error and thus improves quality. This is evident in automated design checking procedures for everything from code compliance (examples include structural details and size of exit stairways) to lighting levels.

Benefits enhancing design quality are considerably less tangible and measurable than benefits increasing productivity and reducing elapsed time. They accrue in the form of a higher level of occupant satisfaction and functional efficiency, productivity, and safety.

Reducing Errors

Financial and operational management—as well as design and documentation—of buildings involves making a large number of individual decisions; coordinating the work of many different people, departments, and organizations; and producing a great deal of highly detailed information. Even in the best-run organization, errors occur with statistical regularity and can result in financial loss, less-than-optimal decisions, occupant dissatisfaction, repeated undertakings of the same work, lawsuits, diverted management time, and high insurance rates. Reducing errors is a very important potential benefit of computer applications to facility management.

Enhancing Management Effectiveness

Very significant benefits can follow from exploiting computer systems to increase the effectiveness of project management, facility management, and general management by means of decision support systems (DSS). These are information gathering and display tools, which usually consist of a general computerized database management system (DBMS) plus management graphic display tools (line, pie, and bar charts, etc.) with display of the information oriented to answering some specific question or aiding some specific decision. Optimally, a decision support system is designed to input questions in English-like sentences. For example, in response to the question, "Which building has most available lease space?" such a system would display a bar chart of the requested data, sorted in descending order. The following purposes are served by means of such a procedure:

- Providing management with what information is needed (rather than with what information is available or obtainable)
- Displaying it in graphic forms (such as color line, pie, and bar charts) that support informed decision-making
- Establishing a more structured and controlled information flow
- Improving the monitoring of resource expenditures
- Achieving greater budget and schedule predictability by replacing relatively unpredictable human performance with more predictable machine performance
- Permitting "what if?" questions, simulation, and forecasting
- Tightening control over information access and security

Regarding the last item in the above list, modern data processing techniques permit much more specialized and selective access to information than was previously possible. For example, today's database management systems permit security codes to be assigned at the level of each individual field in a data record. This means that individual users can be given access to selectively defined pieces of information (for example, the "status grade," but not the salary, of an employee from the personnel file). With a manual system, such safeguards would be much more difficult to create.

Smoothing out Peaks and Valleys

Many design and management organizations experience workload peaks and valleys, to which staffing in the organization must be adjusted. Hence, it is difficult to maintain staff and (especially) workload continuity.

Where computer methods are in use, however, this effect can be smoothed out somewhat: staff that are not working on production can be reassigned to database development, information gathering, or procedures implementation. The benefits gained from these activities show up in the increased scope and higher quality of management decisions later on (when rapid workload increases occur, and the information and procedures already in place are used to handle them).

Hidden Benefits

In addition to the direct and mostly quantifiable benefits outlined in the preceding subsections, there are various potential spillover or hidden benefits that should be considered. Two particularly notable hidden benefits are as follows:

- Achieving a generally more logical, systematic, and predictable approach to procedures, processes, and departmental functions (this is related to the process known as *accumulating intellectual capital,* discussed below)
- Gaining greater understanding of the processes, as a result of pursuing the analysis needed for implementation of an integrated system
- Beginning an intensified, ongoing process of considering and improving facility management procedures and methods (for example, improving the management *of* facility management)

Building the Worth of the Organization

Much of an organization's worth consists of intellectual capital—the technical and operational knowledge possessed by its managers and staff members. This asset is difficult to protect and retain: when key staff members leave, their knowledge leaves with them.

Procedures, software, and data developed within an organization, however, encode such knowledge in a more tangible and more preservable form. Thus, one very important and long-term effect of computer use is to build the worth and effectiveness of the organization by accumulating intellectual capital in the following forms:

- Databases
- Software
- Procedures
- Management techniques

John Adams of the Facility Management Institute (FMI) calls this collecting of intellectual capital "creating *organizational memory.*"

Taking Advantage of System Integration

Automating an integrated facility management system improves the speed and accuracy of the system. While speed and accuracy are two very significant benefits, an integrated system manually executed is still quite beneficial. Integrating a facility management system improves the quality of the decisions reached due to the coordination of information sources. Automation of an integrated system will result in the decisions being reached much more quickly.

The level of integration of a facility management computer system can take any of several forms:

1. *Database integration:* In this form of integration, various application programs can access the same information base.
2. *Communications integration:* Here, various locales, departments, offices, and devices are joined together by digital data communications links.
3. *User interface integration:* Various functions are invoked by means of a consistent command language, either from a standard menu, screen forms, or in some other unified way.
4. *Algorithmic integration:* Various problems can be

handled internally by means of the same algorithm. (An *algorithm* is a sequence of steps—a procedure—often embodied in a computer program.) For example, the same consistent space allocation optimization algorithm can be used to handle site allocation, building stacking, and block floor planning.

QUANTIFYING BENEFITS

All of the benefits previously listed can be quantified into the following four categories (listed in increasing order by dollar savings):

• Savings in facilities group labor
• Savings in construction/implementation/move costs
• Savings in space and resource costs
• Savings in facilities operations costs

These four categories are discussed individually and analyzed in terms of cost in the subsections that follow.

Savings in Facilities Group Labor

Reducing direct costs is an immediate and easily estimated benefit. Since facility management is labor-intensive, with some additional overhead expense, the use of a more systematic and structured approach can reduce costs in most labor categories. This reduced labor on each project allows much higher throughput (in terms of both quantity and quality) for the facility group as a whole. These savings are easy to estimate on the basis of workload and industry averages. Reducing direct costs could mean reducing existing personnel, but organizations implementing systematic facilities procedures and tools seldom lay off staff. Instead, they capture and maintain more comprehensive data, and explore alternatives such as doing needed work (to better the organization's operations) that was impossible to do manually. As procedures become better integrated into the management system of the organization, higher-

level, more experienced staff who leave may be replaced by less skilled staff. Also, some tasks previously performed by relatively more expensive outside consultants may be completed by in-house staff.

All of these labor costs may be estimated by a small percentage (in the range from 5 to 25 percent) of the existing cost (salaries plus all overheads, direct and related costs) of the existing facility management staff. The benefits of being able in the future to perform tasks not currently done (such as preparing special reports to answer periodic questions from top management) because of insufficient staff or time may be estimated by the cost (again, salaries plus all related overheads) of the additional staff or consultants that would have been needed to perform the increased activities using present methods and information sources.

Savings in Construction/Implementation/Move Costs

Improved management, planning, and design concepts reduce construction costs related to unnecessary moves. Real-estate holdings, purchased equipment, and inventories can be held closer to projections, reducing costs further. These savings are more difficult to estimate than the labor savings discussed above, but they are also significantly greater in magnitude.

These represent small-proportion (from 5 to 10 percent) savings on large capital budgets, plus savings in operating budgets in areas such as move costs. In 1986, *IFMA Report Two* stated that employee move costs alone approach $200,000 per year for the average 500,000 square-foot organization.*

Savings in Space and Resource Costs

Larger still is the cost-saving potential of even a very small reduction in the space and equipment resources required. An organization

---

*The formula used was as follows: (500,000 ft^2/200 ft^2 (gross) per employee) × $250 per employee move × 30% churn rate = $187,500.

achieves savings in money spent on anticipatory "staging space" or inventoried equipment, just by making use of the smart decisions resulting from a one-time manual inventory. The results of such an inventory are used to match needs with resources. The inventory lists all of the occupancy needs (size and type of space required, over what time, and so on) and resources (size and type of space, availability dates, and lease restrictions, for example). Generally, the results are used quickly to make matches between needs and resources and for relocation and occupancy decisions, but usually are not retained due to the speed with which the information in a manual inventory "ages" (becomes inaccurate).

The savings typically amount to at least 5 percent of total space. Since rent (or the value—replacement/opportunity cost—of owned space) is conservatively estimated at $20 per square foot per year, and since associated services and overheads amount to an additional 50 percent ($10 per square foot per year), this one benefit alone equals $750,000 per year for an organization using 500,000 square feet. (This figure does not even include the additional savings possible from corresponding reductions in inventoried equipment and furniture associated with the saved space.)

Reduced Facility Operations Costs

By far the biggest savings realizable are in the operation of the facilities. It is clear that the productivity of an organization is directly affected by the quality of its facilities. Factors such as adequate working space, appropriate equipment, and logical circulation all affect production capacity. Reducing other operational costs, such as total maintenance cost, average employee down time during moves, and morale, holds the potential for additional benefits.

These savings are almost impossible to measure accurately or to prove definitely, but most management personnel readily admit that they exist to some degree and are achievable. Over the years, a considerable body of literature has been developed that attempts to prove the benefit and magnitude of such things as the effect of office environment on productivity. It is usually prudent to go with the smallest measure or estimate that is accepted by all concerned management (including, if necessary, no benefit whatsoever). A figure of from 0 to 1 percent of total operating costs (including labor) is a place to start.

Another approach to quantifying benefits is to project what would happen if the computer system were not implemented. For example, a detailed analysis might indicate that the realizable savings from the computer system would barely cover the costs of implementation and operation in comparison to maintaining the present system; however, not doing anything might be the more expensive alternative as manpower becomes more expensive, labor becomes more mobile, and regulations become more numerous and complex.

Sample worksheets showing key benefits variables and an example of savings are shown in figures 3-7 and 3-8.

COST/BENEFIT ANALYSES

As we've seen, the benefits obtained from many costs are difficult to quantify. Accurate data are not often available within the organization's financial system; moreover, the benefit—although clearly existing—may be qualitative in nature and only translatable into a dollar value by circuitous means. For example, improved design quality promotes a happier and more functional occupant, but how many dollars is that worth? Despite these drawbacks, cost/benefit analysis still provides the best basic structure for approaching application selections and implementation decisions.

The cost/benefit evaluation processes that are discussed in this section were developed in the consulting practice of the Computer-Aided Design Group, a firm that for the past ten years

Reduced Labor

Staff Position	Number	Average Cost
Clerical	_____	_____
Professional	_____	_____
Managerial	_____	_____

Use of Outside Staff (Yearly)

Total costs	_____
Average hourly cost	_____
Number of projects	_____

Reduced Implementation Costs

Churn Rate

Percentage	_____
Total square footage	_____
Number of people affected	_____

Number of Projects by Size and Type

Size (sq. ft.)	Number	Average Cost (or Range)
<250	_____	_____
250–1,000	_____	_____
1,000–5,000	_____	_____
<5,000	_____	_____

Annual Construction Budget _____

Annual Furniture and Equipment Budget _____

Reduced Operating Costs

TYPE AND AMOUNT OF SPACE

Type	Amount (sq. ft.)	% Vacant or Misassigned
Office	_____	_____
Lab	_____	_____
Light manufacturing	_____	_____
Heavy manufacturing	_____	_____
_____	_____	_____
_____	_____	_____

Figure 3-7. Worksheet for key benefits variables. (Courtesy of Computer-Aided Design Group)

Reduced Group Labor (in house)

Inventory	0.5	FTE
Req. prog	0.3	FTE
Layout planner	0.1	FTE
Master planner	0.3	FTE
CADD intgr	0.1	FTE

1.3 @ $33.6K = $44K/yr

(Avoid additional person in third yr = $34K/yr)

Reduced Group Labor (outside)

4 to 6 projects @200hrs/project @$50/hr = $40K to 60K/yr

Reduced Implementation Costs

Construction costs	5%	$25K/yr
Move costs	5%	$13K/yr

Reduced Operating Costs

Vacant space (space holdings)	1%	$150K
Down time of employees moved	5%	$10K

Total Yearly Benefits = $282K

Figure 3-8. Savings example. (Courtesy of Computer-Aided Design Group)

has advised major space-user organizations (and the architects and other service firms that assist them) with respect to evaluating and acquiring procedural and automation tools. The major steps of such an analysis are as follows:

1. Develop a list of potential facility management functions (applications).
2. Make initial assumptions (operating environment, number of applications, hardware sharing, and so on).
3. Estimate the total benefits arising from each application.
4. Estimate the total costs of each application.
5. Compute the benefit-to-cost ratio for each application.
6. Rank the potential applications in order, by benefit-to-cost ratio.
7. Assume an implementation consisting of the top n applications whose cumulative annualized cost does not exceed the first year's budget.
8. If necessary, repeat the analysis (starting at Step 2) if any assumptions have significantly changed.

Often, before analysts can begin to quantify benefits, they must calculate basic statistics such as those shown in figure 3-9. In addition, these sorts of cost/benefit analyses have a chicken-and-egg nature: total cost estimates imply knowledge of all hardware costs, for example, but hardware is sized for the number and type of applications that will run on it (still unselected at that point in the analysis). Therefore, an iterative process suggests itself, wherein assumptions and estimates are made on each side,

Quantity	Measure
500,000	Square feet
3,000	Employees
170	Square feet per person
32,500	Equipment
16	Square feet per equipment
4,500	Spaces
112	Square foot per space
8–10%	Growth/reduction rate
25%	Churn rate
125,000	Square feet
750	Number of people in churn
16	Hours of downtime per move
12,000	Total downtime
$15.00	Average hourly employee cost
$180,000	Annual employee downtime
8	Major projects per year
240	Minor projects per year
5–13	Facility group staff size
$33,600	Average yearly staff rates (+33% benefits)
$22,000	Clerical
$36,000	Professional
$52,500	Managerial
$4.00	Per-square-foot move costs
$500,000	Yearly construction budget
$30.00	Per-square-foot space costs

Based on statistics from the International Facility Management Association and the consulting experience of the Computer-Aided Design Group.

Figure 3-9. Characteristics of a typical organization. (Courtesy of Computer-Aided Design Group)

and a solution is converged upon in two (or three or four) passes.

Guidance in developing the applications list can be obtained from several potential sources. Procedures and applications that others in your occupation or industry are currently developing or using should at least make your list—although the dangers of simply copying their efforts (and mistakes) are obvious. Other lists are available from academic courses, professional seminars, and your local IFMA chapter. Several types of consultants are available to perform more extensive evaluations of your organization's characteristics on the basis of the implied list from those special attributes. These include the major accounting/consulting firms and consultants specializing in data processing or even in facility management itself. Vendors, too, suggest potential applications that can be supported by their products. Figure 3-10 shows one example list of possible computer tools.

Figure 3-11 shows an example of an analysis done by the firm of Steinmann, Grayson, Smylie to determine the relative costs and benefits of several computer applications (*modules*) for a client.

Return on Investment

Return on investment (*ROI*) represents the total amount of money or benefit obtained as a result of a given investment of money (cost). It is

COMPUTER TOOL	BENEFITS		
	DESCRIPTION	SIZE %	SOURCE
Inventory	Labor		
	data collection	10	design firms
	in report production	50	design firms
	update edits	20-40	inventory company
	early design labor (tracking vacant)	10	design firms
	Implementation (combined with required program)		
	reduced moves	5-15	consulting
	reduced construction	5-25	consulting
Requirements	Data collection	30	design firms
Programming	Report production	60-80	design firms
	Editing	60-80	design firms
Master Planning	Labor		
	initial alternatives	20-40	estimates
	Development	50-70	estimates
Location and Layout	Labor		
	initial	50	design firms
	revisions	80	design firms
	Operating (all above modules combined)		
	Reduced vacant space	1-5	Consulting

COMPUTER TOOL	BENEFITS		
	DESCRIPTION	SIZE	SOURCE
Floorplan layout			
Document production			
Standards development			
Real-estate acquisition			
Purchasing			
Construction management			
Move coordination			
Cost accounting			
Maintenance			
Budgeting			
Cost estimating			
Project/resource scheduling			
Project/cost accounting			

Figure 3-10. List of possible computer tools. (Courtesy of Computer-Aided Design Group)

usually expressed in terms of an annual percentage. For present purposes, the focus is on expressing the return an organization gets for its investment in a facility management system.

In very simple terms, this may be expressed as follows, using total annualized (life-cycle) costs and benefits derived as has been specified in preceding sections:

Component Costs and Benefits

Module	COST ESTIMATE		PROBABLE BENEFITS							
	Purchase	Develop	Improve Decision	Reduce Time	Reduce Staff	Save Space	Save Furnishings	Reduce Changes	Lower Costs	Benefit Index
1. Projection factors	—	$20,000	•					•	•	3
2. Area projection	$10,000	$40,000	•	•	•	•			•	5
3. Area allocation	$ 5,000	$30,000	•	•		•		•	•	5
4. Analysis	—	$80,000	•						•	2
5. Inventory	$25,000 (partial)	$45,000	•	•	•	•	•		•	6
6. Change management	—	$60,000	•	•						2
7. Financial management	—	$100,000	•	•	•				•	4
8. Drawing system	$40,000–$300,000	$200,000		•	•					2
9. Database	—	included	•	•	•	•	•	•	•	3

Figure 3-11. Component costs and benefits. (Courtesy of Steinmann, Grayson, Smylie)

ROI (%) =

$$\left(\frac{\text{Total annual benefit} - \text{Total annual cost}}{\text{Total annualized cost}} \right) \times 100$$

The first year's ROI is

ROI (%) =

$$\left(\frac{\text{First-year benefits} - \text{First-year costs}}{\text{First-year costs}} \right) \times 100$$

FOUR

SUBSYSTEMS

Facility management is a multidepartmental discipline. For this reason, a facility management system should be viewed as a series of interrelated subsystems (or application modules) that operate in unison. Two broad functional areas should be incorporated within a comprehensive facility management system:

• Information subsystems (decision tools)
• Implementation subsystems (execution tools)

INFORMATION SUBSYSTEMS

Inventories

Two basic inventories are required: an inventory of resources, and an inventory of needs. The resources inventory may be thought of as a simple or annotated land, building, and equipment list and a lease/option list. The needs inventory may be thought of as the architectural program or space needs program. Both inventories require that data be maintained and projected over time. Both inventories also share a common makeup. Three broad types of information need to be maintained in the inventories: personnel information, equipment information, and support space information. Virtually every space requirement can be identified either as one of these three types of information or as a means of directly supporting one of them.

The needs inventory often identifies activities (the functions that need to occur within a space) as primary space consumers. The resources inventory often enumerates leases/options as a way of looking at the spaces in the inventory over time.

Facility types influence the relative presence and importance of people, furniture/equipment, and space as primary space consumers. The function of a building influences the relative amount and dominance of space devoted to personnel workstations, pieces of furniture and equipment, and supporting area. While virtually every facility has all three, the activities that the facility houses (or, in other words, the organizationally driving entity) determine the amount and importance of each.

Office facilities are excellent examples of buildings primarily determined by personnel workstations. While space and services must be provided for office equipment (such as copiers) and support space (such as conference rooms), the major factor influencing organization, space, and services provided is the personnel workstation.

Industrial and manufacturing facilities are excellent examples of buildings primarily determined by equipment needs. Space and services are provided for supporting personnel (such as operators) and areas (such as warehousing), but in this instance the major organizational factors and the biggest influence on space and services provided are the requirements of the equipment used to manufacture the facility's product.

Hotels (and, to a lesser extent, schools) are good examples of buildings primarily determined by space (capacity). Space and infrastructure services are of course also provided for personnel (such as administrative office workstations) and equipment (such as kitchen equipment) but here the primary organizational influence is exerted by exigencies of space for supporting specified activities such as hotel rooms, classrooms, lobbies, and assembly and meeting areas. Another way of describing facility types that are organizationally driven by support space is to say that their needs are determined more directly by the activities that occur within a space, rather than by the people or equipment housed and supported.

Almost all facilities contain needs for people, equipment, and supporting space. The differences lie in organizational dominance—that is, in what supports what. A piece of equipment can support a person (as in an office) or vice versa (as in an industrial manufacturing plant).

The needs and resources inventories not only should contain the physical requirements and availabilities, but also the economic and temporal needs (resources over time). Traditionally, budgets and time constraints/availabilities are recorded separately from enumerations of people, equipment, and space needs and resources; but in an automated system, they should be maintained over time together. Thus, for any point in time, a good inventory contains both needs and potentially available resources for people, equipment, support space, attributes, time, and money.

Allocation

Once the facility manager knows what is needed (the needs inventory) and what is possible (the resources inventory) at multiple points in time, the manager's focus shifts to finding solutions to problems (matches).

A problem arises whenever needs and resources do not match at a particular point in time. (If the two inventories were always the same, there would be no need for facility management.) The allocation function, simply put, consists of allocating resources to fulfill needs: matching each potentially available resource (for personnel, equipment, and support space, taking into account time and money) to each corresponding need, over time.

Ideally, this allocation is optimized; that is, it provides the best fit according to the priorities of the organization, so that the needs are best satisfied. (The term "optimized" here refers to a mathematically provable, best possible solution, and one obtainable within practical constraints.) This means that one of two conditions exist: either all needs have been fully met (and the problem is underconstrained) or all needs cannot be fully met (the problem is overconstrained due to inherent conflicts or logical inconsistencies in the needs). In the latter case, the best solution is the one that allows the greatest number of the most important needs to be satisfied. The terms "greatest number" and "most important" are subject to interpretation (the latter especially so), and a rule structure or weighting system needs to be defined.

Under ideal circumstances, the allocation is optimized over time. This means not only that the greatest number (and most important) of the needs were met, but that they were met by a least-cost path (a path involving the minimum number of lower cost moves).

Traditionally, facility managers were happy if even a majority of the most important needs could be met over time. The number of potential matches between the many specific needs and resources is very large. A major corporation or institution can easily be faced with several hundred or even several thousand potentially viable alternatives, out of a universe of billions of possible combinations. The rules of thumb used by talented facility managers and space planners to attack this problem are complex. Only recently have computer algorithms begun to solve these problems well. In the very recent past, automation has greatly improved both the quality of the allocation solutions and facility managers' confidence in them.

Optimized allocation has the potential to benefit use and operations costs (the major costs of any facility) significantly. Jim Steinmann suggests that the current present-value life-cycle cost of providing office space exceeds $33,000 per employee.* In many instances, it exceeds $50,000 per employee. Even a small percentage increase in overall space utilization efficiency can yield a return considerably higher than the cost of the additional time invested in preparing more comprehensive analyses, space plans, and interior designs.

A company whose space inventory is approaching 1 million square feet and is expected to continue to increase in the future has a major opportunity for savings through improving its space utilization and interior development flexibility. For every 1 percent improvement in space utilization, the space user potentially reduces present-value life-cycle facility occupancy costs by between $1 and $2 million.

Opportunities for improvement are multitudinous. Space costs of more than $33,000 per employee can be reduced through careful response to personnel requirements and—for the space that must be provided—development in the most flexible, cost-effective, and space-conserving manner possible. The experience of consultants such as The Computer-Aided Design Group and Price Waterhouse in other facility-planning exercises indicates that overall im-

provements of between 5 and 15 percent can usually be achieved.

Optimized allocation takes the form of location planning (country, city, site/campus, building, and so on), stacking planning (vertical floor selection within a multistory structure), block planning (layout of schematic departmental blocks on a floor), and space planning (detailed interior layout floor plans), all in consideration of the least-cost move path from a present match of needs and resources to a future match of needs and resources.

Equipment-driven (industrial) facilities have used computer-based mathematical allocation, optimization, and simulation for at least two decades. Indeed, much of the rich body of technical literature supporting optimized space allocation comes out of the professional discipline of industrial engineering (IE). Personnel-driven (office) facilities have only now begun to utilize these techniques. Space-driven facilities (hotels, schools, prisons, and so on) will probably be still slower to use these tools for facility management, given the more fixed nature of these facilities, although they probably will use the tools increasingly for facility design.

According to Charles Reeder, the fundamental inventory and allocation modules for a facility space management system at minimum should provide the following three features:

- *Asset management,* including oversight of all owned and leased property, its current status, its occupants, its allocations, its costs, its functions, and any made or anticipated changes.
- *Reporting mechanisms* to provide decision-support and staff-support information, as required by management, financial, facilities, administrative, and technical personnel
- *Database maintenance,* including generation of historical summaries as data are updated, which

*Steinmann's computation proceeds as follows: (150 ft^2 (net) per person) $\times$ ($80 per ft^2 construction cost) $\times$ (120% conversion to gross ft^2) $\times$ (130% conversion to total development costs) + ($3,500 allowance for furniture and equipment) = $22,220; additionally, (15 years' operating cost of $5 per ft^2 per year for energy, maintenance, repair, and rearrangement) $\times$ (9% discount rate) $\times$ (6% inflation rate) = $11,185; finally, $22,220 + $11,185 = $33,405.

in turn support functions of forecasting, analysis, and evaluation of past performance

IMPLEMENTATION SUBSYSTEMS

Once allocation decisions are made, the focus shifts to physical design and project management. Estimating costs, budgeting, scheduling, drawing, coordinating consultants, and administering contracts are all activities in this subsystem.

Estimating

Preliminary budget estimates define the economic limits on design. *Cost estimating* is a process of identifying and associating unit costs with unit quantities derived from the needs inventory; typically, it is an interactive process of gaming alternative selections until a compromise satisfactory to both the client/user and the designer is reached between cost and quality/functionality. Budgeting is then a process of formalizing the selection into dollar limits.

Design

Physical design consists of a number of components. Space planning and interior design are processes involving the detailed design, selection, and layout of components. Working drawings and specifications are produced as the primary contract documents for directing the contractors who will build and the installers who will assemble the building materials, furniture, and equipment specified.

Administration

Contract administration consists of observing and supervising the work as it is performed. Scheduling is a process of observing scheduled milestones and activities, replacing estimated dates with real ones as the work progresses, and adjusting various resulting and dependent activities accordingly.

Often techniques such as CPM (critical path method) or PERT (program evaluation and review technique) are utilized. These are established project management techniques that assist in determining and managing the precedence of tasks required to complete a project. They are well documented in the literature. One good treatment is Bennett 1977.

Operations

Ongoing facility management operations consist of all the implementation activities discussed here, plus economic, maintenance, and general decision-support activities.

Economic Management

Economic management comprises all attempts to manage operating resources and capital resources (operating and capital budgets). The subsystems here are financial reporting tools.

Operating resources management subsystems include tools designed for such areas as the following:

- *Rent rolls*
- *Lease files* for situations in which the manager is lessor and/or lessee
- All *associated operating expenses*, potentially including areas as diverse as parking, security, and taxes
- Accurate *cost accounting* and *chargeback* schemes

Chargeback refers to the accounting methods used to charge users/occupants for facilities resources. The principle resource charged-back for is space. A majority of organizations have chargeback schemes for at least the rent component of space costs. It is less common to have chargeback schemes which include all space occupancy costs (such as security, utilities, and services) or equipment costs.

Capital resources management subsystems include information on depreciation, present value and market value of land and fixed assets such as plant (buildings), equipment, furniture, and fixtures.

For both operating and capital resources management, money management tools are used, particularly those that track preformance to a budget (either an operating expenses or capital expenditures budget or both).

Maintenance

Maintenance systems provide a management structure for periodic servicing of equipment, preventive maintenance of equipment, and the scheduling of routine building service to replace light bulbs, paint, and so on. Maintenance subsystems include periodic/preventive maintenance and unscheduled maintenance. The goals of the subsystem usually are to eliminate the unscheduled maintenance and (within that) to minimize the frequency of preventive maintenance.

Decision Support

Given the preceding information on economic management and maintenance, sophisticated decision-support systems for the facility manager become possible, including systems for assessing lease versus build versus buy options and similar decision information.

As has been noted continually, most of the information required by the facility manager must be maintained for more than one simultaneous moment of time. Most of the database specifies things that occur or change over time. Thus, inventories, optimizations, and other data increase significantly in value to the manager when they represent past history, present status, and future plans and projections.

FIVE

DATABASE MANAGEMENT

A facility management system requires an extensive base of detailed information. The process of developing and maintaining an accurate and current set of quantitative and qualitative information in machine-readable form is termed *database management*. A database management process must be capable of performing the following functions:

- *Assembling* large amounts of information
- *Storing* that information in a manner that is accessible (retrievable and updatable)
- Fluidly permitting *sorting, reporting,* and *interactive querying*
- Encompassing procedures for *maintaining* accurate and current information

A facility management database should contain information about physical space resources (the space/occupancy inventory); historical, current, and projected needs (staff, equipment, and space requirements); fixed asset (furniture and equipment) inventories; cost and schedule information; inventories of leases and excess property, furniture, and other standards; and a variety of associated (attribute) information.

Computers obviously are ideal for database management. Information should be easily retrievable by a variety of users, but at the same time safety and security issues (such as who

may read and write various items of information, on what authority, and with what "audit trail" record) need to be addressed. Some access limitations and authorization procedures to the database system generally and to individual portions of it (files, records, fields, information regarding specific departments or users, and so on) should be developed.

Facilities decisions are only as good as the information on which they are based. Any active organization's facilities unfortunately present a moving target: any comprehensive inventory performed from scratch is likely to be out of date by the time it is completed. If, however, the database of facts and projections is conscientiously updated as soon as changes occur (that is, as soon as they are detected by procedures), the window of uncertainty is considerably narrowed. Instituting dynamic update procedures to shorten the time between occurrence and discovery of change heightens accuracy still more.

It bears repeating that effective procedures, not automation per se, distinguish successful facility management systems from unsuccessful ones. Computers positively and mightily affect the speed, accuracy, and economy of the facility management effort—but only those variables. Computers will make a logically organized facility management department function faster, more accurately, and at lower cost; they will help a disorganized facility management operation only in processing inaccurate information and producing error faster and cheaper.

Charles Reeder was quoted in a computer trade publication as saying, "If you computerize a mess, all you get is a computerized mess...." This clearly is true.

DETAIL

The amount of detail elicited for describing facilities should not necessarily be the greatest amount possible. The first thing to determine is what output (information) is desired from the system. Desired detail of output (including anticipated future desires) is the best determinant of input.

Provision should be made for automatically archiving databases and inventories at regular intervals. These historical facilities records are a valuable resource for determining past patterns and for projecting future trends.

ASSEMBLING THE DATABASE

Assembling the wealth of information required in a facility management system database for a large organization is time-consuming and potentially costly. Information is needed on land and buildings; space, personnel, and equipment; historical trends; marketing and financial data; and so on. The information assembled must be comprehensive and should reflect at least a few years of organizational experience. The data must be gathered accurately, inputted into the database, checked for errors (both of content and of entry), and verified. The only thing worse than having no information is having inaccurate, out-of-date, or incomplete information—the manager without a database at least knows what he or she does not know.

Perhaps the single most important responsibility of the facility manager is to maintain an accurate and up-to-date (data)base of in-

formation. With the information in hand, management can make decisions based on facts, and facility managers can use informed judgment to develop future projections, analyze alternatives and planning philosophies, develop master plans, and implement specific projects. Having a consistent and comprehensive statistical database also ensures uniformity (in terms of budgets, standards, reporting conventions, and so on) from department to department as well as over time.

Peer Databases and Bridging

It is critical that the facility manager be aware that most (60 to 100 percent) of the information needed for the organization's facility management database already exists within the organization. (In fact, in most large organizations, it already exists—somewhere—in machine-readable form.) Examples of peer databases include fixed assets, corporate or strategic planning, CADD, accounting, and personnel.

The biggest variable in the cost-effectiveness of keeping the facility management database current is how successful the facility manager is in locating and utilizing the information from peer databases. Often this requires political skill. The facility manager needs to convince other managers within the organization that giving him or her access to their databases (with any appropriate limits imposed as security measures) is in their (and the organization's) best interest.

Usually the facility manager's task also requires data processing skill in order to build interfaces (or bridges). Many consultants are available to assist with this effort, and some standard interfaces (such as the Computer-Aided Design Group's Facility Drafting Coordinator) are beginning to appear in the marketplace. Invariably there is a requirement for procedural thought on such issues as how and when to perform updates. A good approach to this question is the same one used when the level of detail in the database is to be determined: examine the desired output.

AUTOMATION

Facility management is information management. Clearly the capacity of computers to manage large amounts of data accurately and inexpensively makes their application to facility management systems very attractive. Because many of the operations required in facility management involve projections and inventories, major time savings are possible. Among the fundamental advantages of computer use are speed, accuracy, consistency, and cost. Information can be conveniently compared in various ways, affording faster and more complete analysis.

Data Elements

A facility management system requires a multitude of data elements. Following is a listing of some types of essential information that may serve as the beginning of an information-gathering effort. The list is by no means exhaustive. The examples have been restricted primarily to inventories, since they constitute the preponderance of the facility management database. Other applications—project management, allocation (stacking and blocking), and so on—access these inventories, as well as make use of additional small (and sometimes temporary) databases of their own.

Space: Space information is expressed in net square feet (NSF), gross square feet (GSF), rentable square feet (RSF), and location (country, region, site, building, floor, room, and so on). Data on square footages are needed, by building and by organizational group (division, department, and so on), for all owned and leased space. Building core and common areas are deducted from GSF to yield NSF. RSF, as given in lease agreements, should also be available.

Type: Type of structure is expressed in terms of uses supported or possible (warehouse, office, laboratory, auditorium, and so on), location, and special footnotes and attributes regarding the space (infrastructure or services available, "highest and best use," and so on) that may define its limitations.

Quality: Included in qualitative information are the condition of the structure, its compatibility with building codes, and descriptions of conditions that may suggest or preclude specific uses (such as there being no off-street parking available, there being backup power generation available, and so on).

Use and Allocation: Present use of all space, by department and function (private office, storage room, and so on), is the primary focus of use and allocation information. The data should identify number of users (for example, nine-person conference room), type of space (for example, workstation standard XYZ-6), and so on.

Space Assignment: As the heading makes clear, the information here identifies the occupants to whom and the activities to which the space is assigned. In many organizations, the data must cover a multitude of other assignees (the owner may differ from the person or entity to whom the space is billed, as well as from the occupant, the cost or profit center, the project, the customer or grantor, and so on).

Occupancy: Occupancy-related information includes the numbers and identifiers of people, equipment, and activities occupying each space (office, workstation, warehouse, print shop, and so on.) In addition, it covers calculations of net area factor (NAF, obtained by dividing NSF by total personnel supported) vacancies, and the like. Names, titles, and organizational affiliations are also needed.

Staff: Staffing data are required by group (such as division or department) and group total, plus total by building, floor, city, room, or other specified area. Area factors can be net (as in

NAF, discussed above), gross, rentable, and so on.

Area Factor: *Area factor* is defined as area per person. This quantity generally must be computed by department, division, group, building, floor, and overall total.

Operating Costs: Information required under this heading relates to costs such as those for security, utilities, remodeling/construction amortization, moving, and cleaning (each calculated by department, floor, building, and total). It also includes a tracking of rate of change over time, and data for any location by occupancy (lease rates), for maintenance and/or location-specific operating costs, and for other fixed costs (such as taxes).

Furniture Inventory: This inventory may list furniture (desks, chairs, files, acoustic panels, credenzas, and so on) by location, assigned department, manufacturer, color, size, condition, associated purchase order, depreciation method, and other such variables.

Equipment Inventory: This inventory lists office and/or manufacturing equipment by location, assigned department, manufacturer, color, size, condition, associated purchase order, depreciation method, or other attribute (for example, required electricity, cooling, piping, access, noise, emissions, or all-floor loading).

Lease Register: In general, a lease register includes data on a unique identifier (lease number), amount of space, location, occupancy by department, cost, rate(s) of increase(s) over time, renewal options, expiration date(s), financial and other unique/key terms of lease.

Historic Trend Data: This information consists of summary data per year (for perhaps the past ten years) or other time period for staffing, costs of space, and square footage occupied per department. Data relevant to rate of change (such as production and sales) should also be included.

Forecast Data, Business Plan: Planned/ anticipated indicators of business change are covered under this heading. These may reflect changing space quantity, location, or type; hence, projected number of sales, records kept, customers processed, and projects completed should be recorded. The category may also include such data as locations of clients, increases in the number of drafting positions while the number of typists decreases, the effect of modernization (more or less staff or space), changing utility needs, and so on.

Demographics: Principally, demographic data are of interest in planning or management involving employees, but occasionally market area information is useful for production projections. A demographic profile may include factors such as income, residence location, distance of commute, number of male and female residents, their age, other information potentially affecting consideration of new building locations, number and size of restrooms in each, size of workstation versus level of status/achievement, and personnel duties and responsibilities.

Area Blockouts: These consist of graphic depictions (roughs "as built") in plan and profile of space as allocated to each department. Their purpose is to reveal relationships between departments horizontally (block plans) and vertically (stacking plans), with accompanying data on square footage.

As-Built Plans: These consist of recorded working drawings (accurate, detailed plans) of each floor, with type and location of furniture, workstations, and equipment. Scale drawings of any floor should be provided at a uniform scale or in specific detail (denoting departments).

INVENTORY CATEGORIES

Both resources inventories and needs inventories are composed of equipment, space, and personnel inventories expressed over time.

Equipment and Furniture Inventories

Large organizations own many thousands of pieces of furniture and equipment. It is often advisable to make an inventory of these before making a planning decision. An organization typically reuses some equipment and furniture, discards others, and purchases additional new items.

The forms in which databases can be maintained range from brief nongraphic records for each item to color photographs and complete descriptions of the fixture, its materials, finishes, costs, supplier(s), requirements, relationship with and dependence on other items in the inventory, and a computer graphic representation of the item. Tracking items in the database can be accomplished with each item assigned to a group, subgroup, or person. Records within the database can be associated with the physical objects by using a machine-readable bar coding system.

Trained staff members or a space-planning firm usually carries out the inventory process, using formatted worksheets (examples are shown in figures 5-1 and 5-2). Data can then be entered directly from a keyboard and linked to additional information such as a photograph or catalog sheet. Nonetheless, considerable difficulty is involved in carrying out the inventory and in developing and maintaining a large and detailed database. Figures 5-3 and 5-4 illustrate typical inventory reports.

It is virtually always necessary to update an inventory database at intervals. Good database management systems provide convenient, English-like language statements for specifying the updates that are to be made.

SQL (Structured Query Language) is one industry-standard way of achieving this. While SQL is not yet fully standardized, most SQL-supporting commercial DBMS vendors will likely modify their products to conform to the standards as they solidify. (Readers new to data processing should be aware that *SQL* can be both a product name—such as IBM's SQL/DS—

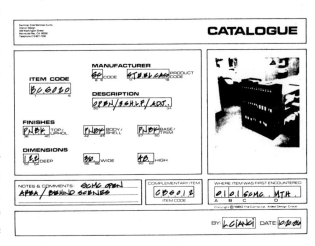

Figure 5-1. Catalog form. (Courtesy of Cole Martinez Curtis and Associates)

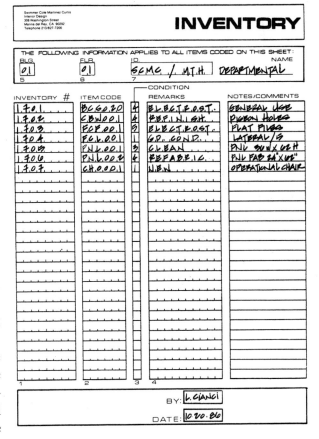

Figure 5-2. Inventory form. (Courtesy of Cole Martinez Curtis and Associates)

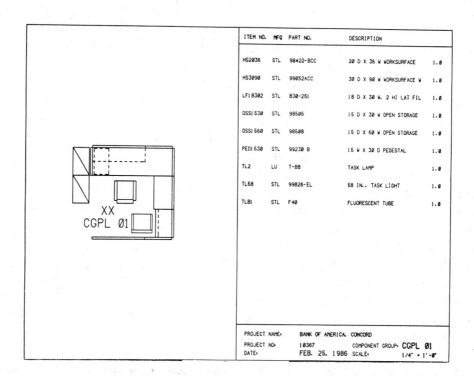

INVNTORY NUMBER	ITEM CODE	M F	PRODUCT CODE	DESCRIPTION	FINISHES UPHLS SHEL TRIM	DIMENSION W D H N	C BL FL	CURRENT CLIENT CNSLT	BL FL	FUTURE CLIENT ROOM#	Q	NAME/TITLE REMARKS	LAST UPDATE
01	PLA005			FLOOR		132 132 142	01 03	19 RES DEPT	XX				
0016001	FCX005			CARD CATALOG	WVOM WV	PL8K 403 019 593	3 01 04	2 EXPL OR	XX			REFIN	850418
0016002	FCM001	HS	HARPERS		PNCM PNCM	PSCR 042 183 060	2 01 01	582	P			GD COND	850418
0147	FCG073	AM		VERT.4/LEGL.KEY	WVWDQ WVWD	ATBS 018 242 513	2 01 03	14	P			GD COND	850418
02710	CHA013			SECY/TLT SWVL/C	VNBK VNBK	BSAL 017 020 033	4 01 02	2E	XX			ELEC.STAT.	850417
1-477	CHQ028	UN		GUES.4-LEG	LTBN LTBN	WVWM 022 021 037	2 01 09	38 MGR OPER	E			GD.COND.	850417
100676	FCG009	AM	ART METL	VERT.4/LEGL KEY	PNCM PNCM	BSCR 018 242 513	3 01 06	2H EXP LOR	XX			ELEC.STAT.	850418
100760	FCG058	UN		VERT.4/LEGAL.K.	PNCM PNCM	BSCR 018 028 503	3 01 06	2H EXP LOR	XX			ELEC.STAT.	850418
1009972	FCB003			VERT.3/LETTER	PNCM PNCM	PNCM 015 281 042	2 01 03	TEXACO	XX			GD COND	850418
1009974	FCB003			VERT.3/LETTER	PNCM PNCM	PNCM 015 281 042	2 01 03	TEXACO	XX			GD COND	850418
10126	FCG027			VERT.4/LGL.KEY	PNTN PNTN	ATBS 018 262 053	3 01 01	582	XX			ELEC STAT	850418
10128	FCG027			VERT.4/LGL.KEY	PNTN PNTN	ATBS 018 262 053	3 01 01	592	XX			ELEC STAT	850418
10129	FCG027			VERT.4/LGL.KEY	PNTN PNTN	ATBS 018 262 053	3 01 01	582	XX			ELEC STAT	850418
1019	FCG006			VERT.4/LEGL.KEY	PNBNQ PNBN	ATBS 171 241 512	3 01 12	13AA FILE	P			VERT LEGL.	850418
102557	E9Z002	UN		COAT RACK	WVWL WVWL	ATBS 022 022 072	3 01 09	3C	E			REFIN	850418
102563	FCG050		STEELCAS	VERT.LEGL/4 KEY	PNTN PNTN	PSCR 173 282 521	3 01 07	4 COMP TROLR	XX			ELEC.STAT.	850418
102565	FCG050		STEELCAS	VERT.LEGL/4 KEY	PNTN PNTN	PSCR 173 282 521	3 01 07	4 COMP TROLR	XX			ELEC.STAT.	850418
102616	FCH029	SC	STEELCAS	VERT.LEGL/5 KEY	PNTN PNTN	PSCR 173 282 582	4 01 03	16A	XX			ELEC STAT	850418
102617	FCH053	SC	STLCASE	VERT.5/LEGAL.K.	PNCM PNCM	BSCR 018 282 581	2 01 04	16A	P	04	475	GD.COND.	850916
102618	FCH053	SC	STLCASE	VERT.5/LEGAL.K.	PNCM PNCM	BSCR 018 282 581	2 01 04	16A	P	04	475	GD.COND.	850916
102619	FCH053	SC	STLCASE	VERT.5/LEGAL.K.	PNCM PNCM	BSCR 018 282 581	2 01 03	16A	P	04	475	GD.COND.	850916
102643	FCH053	SC	STLCASE	VERT.5/LEGAL.K.	PNCM PNCM	BSCR 018 282 581	2 01 04	16A	P			GD.COND.	850916
102644	FCH053	SC	STLCASE	VERT.5/LEGAL.K.	PNCM PNCM	BSCR 018 282 581	2 01 04	16A	P	04	475	GD.COND.	850916
102645	FCH053	SC	STLCASE	VERT.5/LEGAL.K.	PNCM PNCM	BSCR 018 282 581	2 01 04	2 EXPL OR	P			GD.COND.	850418
102646	FCH053	SC	STLCASE	VERT.5/LEGAL.K.	PNCM PNCM	BSCR 018 282 581	2 01 04	16A	P			GD.COND.	850916
102647	FCH053	SC	STLCASE	VERT.5/LEGAL.K.	PNCM PNCM	BSCR 018 282 581	2 01 04	16A	P	04	475	GD.COND.	850916
102673	CHA013			SECY/TLT SWVL/C	VNBK VNBK	BSAL 017 020 033	4 01 05	2 EXPL OR	XX			ELEC.STAT.	850417
102691	DKL003	UN		TRAD.FLUSH	WVWM WVWM	WVWM 085 045 030	2 01 12	13AA ASMGR	E			TRAD	850418
102694	CHQ015			GUES.4.LEG	LTBN LTBN	WVWM 241 201 035	2 01 04	16A	E				850417
102697	TBB003	UN		SIDE/RD.FOURLEG	WVWM WVWM	WVWM 034 017 030	2 01 12	13AA ASMGR	E			MTCH CRA03	850418
102716	FCH015	AM	ART METL	VERT.5/LEGL KEY	PNCM PNCM	BSAL 018 282 581	3 01 11	14AA GENRL				ELEC.STAT.	
102717	FCH015	AM	ART METL	VERT.5/LEGL KEY	PNCM PNCM	BSAL 018 282 581	3 01 11	14AA GENRL				ELEC.STAT.	
102732	STA006	UN		DRAFTING	LTBN LTBN	PNBN 017 013 039	3 01 04	2H EXP LOR	P			REUPHOL	850418
102733	FCG009	AM	ART METL	VERT.4/LEGL KEY	PNCM PNCM	BSCR 018 242 513	4 01 05	2 EXPL OR	XX			ELEC.STAT.	850418
102746	FCH052	AM	ART MTL	VERT.5/LEGAL	PNCM PNCM	BSCR 018 282 582	3 01 06	2H EXP LOR	XX	04	475	ELEC.STAT.	850916
102751	FCH029	SC	STEELCAS	VERT.LEGL/5 KEY	PNTN PNTN	PSCR 173 282 582	3 01 06	150PER ATION	XX	04	475	ELEC.STAT.	850916
102759	STA006	UN		DRAFTING	LTBN LTBN	PNBN 017 013 039	4 01 04	2 EXPL OR	P			ELEC.STAT.	850418
102764	DKL023	UN		TRAD.FLUSH	WVWM WVWM	PWBN 652 036 282	2 01 10	14	ATTY	E		GD.COND.	850418
102765	CHG006	UN		DESK.TLT.SWVL/C	LTBN WVWM	ATBS 024 020 373	2 01 11	14AA ATTY	E			GD.COND.	850418
102771	CHQ049	UN		GUES.FOURLEG	PNGN PNGN	PNGN 020 017 032	4 01 05	2 EXPL OR	XX			ELEC.STAT.	550418
102785	CHG008	UN		DESK.TLT/SWVL	LTBN LTBN	WVWM 232 020 039	3 01 11	14AA SRATY	E			GD.COND.	850417
102797	CHF003			DESK TLT SWVL/C	BNGD VNGD	BSAL 022 022 031	4 01 08	4 COMP TROL	XX			REUPHOL	850417
102802	DKL018	UN		TRAD.FLUSH	WVWM WVWM	ATBS 066 036 301	5 01 10	14	ATTY	P		REFIN	850418
102805	CHQ023	UN		GUES.FOURLEG	WVWM WVWM	WVWM 202 182 322	2 01 03	16A	E			GD COND	850417
102806	CHQ023	UN		GUES.FOURLEG	WVWM WVWM	WVWM 202 182 322	2 01 03	16A	E			GD COND	850417
102812	CHQ023	UN		GUES.FOURLEG	WVWM WVWM	WVWM 202 182 322	2 01 02	5A MGR FREMN				GD COND	
102819	CHQ049	UN		GUES.FOURLEG	VNGN PNGN	PNGN 020 017 032	4 01 02	3I	XX			ELEC STAT	850418
102830	FCH048			VERT.LEGL 5/KEY	PNCM PNCM	PNCM 182 282 582	2 01 08	3C	F			GD.COND.	850418
10291	FCG028			VERT.4/LEGL.KEY	PNTN PNTN	ATBS 021 263 053	3 01 09	2C REC ORDS	XX			ELEC.STAT.	850418

Figure 5-3. Inventory report. (Courtesy of Cole Martinez Curtis and Associates)

ITEM NO.	MFG	PART NO.	DESCRIPTION	
HS2036	STL	98422-BCC	20 D X 36 W WORKSURFACE	1.0
HS3090	STL	99052ACC	30 D X 90 W WORKSURFACE W	1.0
LF18302	STL	830-251	18 D X 30 W. 2 HI LAT FIL	1.0
OSS1530	STL	98505	15 D X 30 W OPEN STORAGE	1.0
OSS1660	STL	98508	15 D X 60 W OPEN STORAGE	1.0
PEDI530	STL	99230 B	15 W X 30 D PEDESTAL	1.0
TL2	LU	T-88	TASK LAMP	1.0
TL58	STL	99826-EL	58 IN.. TASK LIGHT	1.0
TLB1	STL	F40	FLUORESCENT TUBE	1.0

XX
CGPL Ø1

PROJECT NAME:	BANK OF AMERICA. CONCORD	
PROJECT NO:	10367	COMPONENT GROUP: CGPL Ø1
DATE:	FEB. 25. 1986	SCALE: 1/4" = 1'-Ø"

Figure 5-4. Inventory report. (Courtesy of Skidmore, Owings & Merrill)

and the query language standard.) Many vendors now produce software products that are subsets or supersets of the (IBM) standard.

Good facility management systems, however, go a step farther than good database management systems do, providing applications and interfaces tailored to the facility management user. Menu-driven screen forms are an example of such interfaces. Such forms make the creation of queries, updates, and database administration functions much simpler, requiring only that the user select actions and data items from a predefined screen-form menu. Figures C-8 and C-9 of the color insert show examples of screen forms. Figure C-10 shows the use of windows that allow the user to display simultaneously alphanumeric screen forms for database work, CADD/graphics, reports, and so on.

Flexible output formats and easy-to-use query languages also can be brought to bear on specific one-time problems. For example, a user can identify items to be moved, repaired, or discarded, and then print moving labels, as illustrated in figure 5-5.

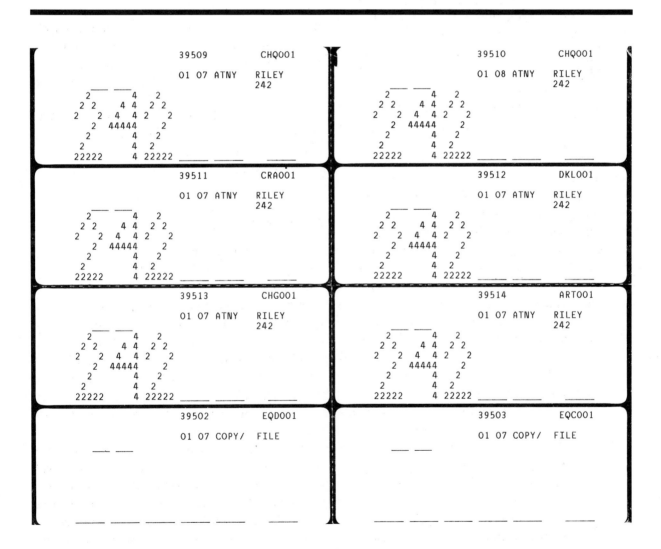

Figure 5-5. Furniture/equipment moving labels. (Courtesy of Cole Martinez Curtis and Associates)

Space Inventories

Space inventory databases maintain a description of the physical framework within which the organization is housed. Depending on the degree of detail and sophistication required, the depth of the description varies from minimal (a description consisting only of square footages) to great (one involving extensive detail, including attributes and relationships, and even three-dimensional graphic representations of the spaces and attached support components).

A space inventory receives general, unstructured descriptions of the organization as input and organizes the records to provide output to other modules of the system. In addition, tabular reports can be made directly from the database of the current space inventory, or the database can be queried for particular items of information (such as how many conference rooms are in a department).

A space inventory can be handled in much the same way as are furniture and equipment inventories. Occupancy, dimensions, areas, finishes, and fittings of spaces are in most cases recorded, together with assessments of the suitability of each for specific architectural functions or occupancy activities. Often a cost of conversion is associated with alternative uses. In addition, aggregations of space into larger groupings must also be noted, since it is often necessary to associate areas and other properties in such a manner. These aggregations can be physical (for example, rooms into floors and thence into buildings, campuses, and so on) or organizational (rooms into departments and thence into divisions, subsidiaries, and so on).

Reports can be outputted in the form of plotted plans, as an alternative to reports in alphanumeric format. This is accomplished by storing in the database a geometric description of each room. The description often consists of simple single-line polygon graphics (enough only to provide block-plan drawings), although occasionally organizations may find it cost-effective to store additional detail (sufficient to produce as-built drawings) by use of or interface with a CADD system.

The space inventory database contains at minimum a physical description of the spatial components (size, location, type, and so on) and listings of related nonphysical attributes (lease information, rentable/usable areas, and so on), including identification of the group to which the component is assigned. For a given building, therefore, the database contains perimeter and/or core descriptions. Individual floors are, where possible, described by reference to these standards.

Detailed space descriptions are coordinated to include the following: necessary support facilities, such as telephone and electrical connections, as well as other services and supply lines; horizontal and vertical circulation patterns; positions of structural and nonstructural elements; and finishes. Although such a description need not be three-dimensional (or even graphic), it can be coordinated with a CADD graphics system. In that case, schematic floor plans are maintained in association with related nongraphic data.

Personnel Inventories

The personnel inventory is the third major inventory component in a facility description. It contains at minimum a list of all employees or members and their positions and/or affiliations within the organization. A full database also associates space requirements, job classification codes, special needs (for example, handicaps), preferences (for example, smoking), status/grade, adjacency/workflow/communication needs, and other particular requirements with the personnel record.

Organization charts and current personnel rosters that classify personnel by name, level, task definition, organizational affiliation, and responsibility should be provided. Both formal and informal working organizational charts should be maintained, together with a functional description of the responsibility of each department, group, or subgroup.

As in the case of the two previous types of inventories, output from the personnel inventory database takes the form either of data to other modules in the system or of sorted tabular reports of current resources. Online querying of the data is possible from all three inventories as an additional output source.

Essential functions of the databases are to develop and handle bulk input of the initial inventory; to maintain, edit, and update the database; and to generate standard reports. Useful subfunctions include generating special ad hoc reports, conducting interactive querying, and performing automated procedures for checking the consistency of records.

Historical Data

Information regarding past staff levels, space requirements, equipment needs and uses, production quantity, and task definitions should be maintained for as many years in the past as are available and useful (up to five or perhaps ten years). Maintaining this summary-level historical data provides the planner with a perspective of operations that makes more accurate assessments of future requirements possible.

Summaries of data for projecting and modeling future scenarios can be provided, as can straightforward reports of the operation's status at points in the past.

Most organizations accumulate historical information randomly. Analysis is necessarily tedious and involves the costly manual archiving and indexing of large volumes of data. With an automated system, the same results can be achieved in much less time simply by recording a summary of the contents of the database at regular intervals and prior to making major changes in the data. Archiving this information regularly can (and should) be an automatic part of the facility manager's procedures. Thus an accurate historical record is maintained, from

which data can be easily drawn to make, check, or validate projections of future growth.

Needs Inventories

A needs inventory (also known by architects as a requirements program) consists of procedures and computer programs that serve to collect and manage information about group space, personnel and equipment needs, communication patterns, and interaction requirements.

Questionnaires and interview forms are the tools most often used to solicit a manager's knowledge of group needs for activities, personnel, equipment, and support spaces—both current and projected. A needs inventory manipulates square-footage data, based on the information gathered. Reports are generated by the computer to provide the user with detailed analyses of each group's square footage, equipment, and adjacency and services needs, as well as various summaries. A needs inventory can be used at a general level for sizing the building and making area projections. At a detailed level, a needs inventory can be used for determining criteria for interior planning.

The needs inventory establishes the space required over time by various organizational groups. The time component consists of plan dates: points in time at which information about space requirements is desired. These are often set to coincide with an organization's expansion and development milestones.

The total area required by a group is determined from the sum of the areas required for the group's personnel, equipment, and support space. This figure can and should be adjusted by means of factors to account for internal circulation, design concepts, building-space characteristics, contingencies, and so on. A wide variety of summary and detailed report formats are desirable with the needs inventory.

A needs inventory database should be able to work with different amounts of data and at different levels of detail, and still produce useful reports. The amount of data supplied may vary according to the amount of detail required in the report or the amount of information available when gathered, either top–down or bottom–up.

In addition to supplying reports on space needs, a needs inventory can report on pieces of information that are not directly concerned with the sizing of the facility. Known as *attributes*, these pieces of information provide a way of easily extending the description of items.

Attributes can be assigned to groups, persons, equipment, and/or support spaces. By this means, for example, equipment having special voltage requirements, facilities providing for handicapped persons, or support spaces having unique HVAC requirements can be located, modified, and/or reported. The needs database is a basic component of any effective facility management system, whether manual or automated.

SIX

SPACE ALLOCATION AND PLANNING

The location and the layout of physical facilities are important factors in the management of the capital investment in buildings. Planning related to location and layout also has a significant impact on the costs associated with the equipment and people occupying buildings.

Typical location and layout problems include the following:

- Selecting a site for a new facility
- Assigning operating groups to cities, sites, campuses, buildings, and so on (location planning)
- Assigning groups to floors within a multistory building (stacking planning)
- Locating equipment and personnel groups on the floors within facilities (block planning)
- Determining the detailed layout of furniture, equipment, and services on floors (space planning)
- Reassigning space in connection with a move

Accommodating future area requirements with an eye to all of the important variables is a complex problem, to which various solutions are possible. Allocating area requirements should fulfill a number of specified purposes:

1. It should satisfy the space needs of all employees and working groups.
2. It should not compromise the productivity of the organization. Ease of communication, material, and information flow should be maximized, strengthening cohesion among groups

that must work together and achieving contiguity of closely interrelated groups.
3. It should minimize fixed operating expenses, such as rent.
4. It should minimize wasted and/or unusable space and equipment.
5. It should anticipate and allow for growth and shrinkage over time.
6. It should take into consideration lease expirations, availability of capital for new construction, and similar matters.
7. It should allow for its application to multiple buildings.
8. It should model alternative plans and management strategies.
9. It should proceed vertically from floor to floor in a building and horizontally on each floor.
10. It should serve to minimize moving costs over time, while tending toward a global optimum.
11. It should maximize responsiveness to the location and allocation preferences of management.

Sophisticated mathematical and operations research techniques exist for allocating required areas within built space, subject to constraints of floor space availability, distances between placed areas, costs of space, costs of moving from one position to another, travel time, required adjacencies between allocated areas, and so on. When formulated as algorithms and data for computer programs, these techniques offer rapid generation of near-optimum alternatives in single or multiple structures. These alternatives can then be reviewed and changed by management, rerun in the computer, and compared to determine optimal future facilities solutions. Simple graphic presentation of the results is possible, using color vertical stacking layouts for multistory buildings, and horizontal block floor plans for each floor.

Software for automated and computer-aided layout and space allocation is becoming available. Such products have attracted great interest from major institutions experiencing significant and continuing space management problems—telecommunication companies, financial institutions, and others whose macroeconomic environment is changing. These facility allocation programs often are used in conjunction with the database in which the organization maintains the inventory of its physical facilities, including its location, condition, occupants, and equipment.

This chapter is organized to explore the space allocation and planning process sequentially. First it discusses the input data to the space planning and allocation process, and methods of projecting these data. Next it describes the space allocation process itself, as well as computer programs that are used to assist in the process. Then it takes up furniture and equipment selection and space planning (detailed layout, CADD, and contract documents). Last it explores postoccupancy activities, system integration, long-range planning, and scope of services.

PROJECTIONS

Projecting space requirements can be done relatively quickly to gain high-level answers, or it can be performed in a detailed fashion for more specific needs. Underlying personnel or process projections and the resulting space projections can be obtained through various approaches, as described in the ensuing subsections.

Projection of Personnel by Annual Growth Rate

An estimate of the number of future personnel in the organization (or at least a range within which future personnel levels are likely to fall) is most readily developed by defining annual rates of growth that can reasonably be expected to occur, based on historical trends and cursory knowledge of the anticipated future environment(s). For example, if an organization's total personnel level has increased during recent years at an average compounded annual rate of 1.5 percent per year, and if the change between any two consecutive years has been an increase of not less than 1 percent and not more than 2.5 percent, it would be reasonable to assume (in the absence of any information to indicate that future growth patterns will change) that staffing levels will continue to increase at an average annual rate of 1.5 percent.

Because the future does involve a certain degree of uncertainty, however, a more conservative approach under the preceding facts would be to assume that future growth will continue at not less than 1 percent annually and not more than 2.5 percent per year. This allows for the fluctuations of past years but also conveys a degree of certainty that growth will continue to occur within its historical range. Thus, by the first approach, if the base personnel level were 10,000 and a 1.5 percent compound average annual growth rate were projected for five years, the projected personnel level after five years would be 10,773. Using the second approach, the future level would be projected to fall somewhere between 10,510 (the 1 percent annual growth rate figure) and 11,314 (the 2.5 percent annual growth rate figure).

Increasing the difference in the assumed annual compound rate of growth by 150 percent in this case (the difference in magnitude between 1 and 2.5 percent), starting with a base of 10,000 employees, indicates a variability in total future personnel levels (and associated space requirements) of only 7.65 percent (the difference between 11,314 and 10,510 employees, divided by 10,510 employees). Divergent ranges of annual growth rates thus can be accommodated.

In the preceding example, the single projection of 10,773 personnel provides a more specific planning target than does the range from 10,510 to 11,314. The degree of confidence in the projections is obviously much higher, however, for the figures obtained using the range approach. When used for general space-planning purposes, projections made by defining a growth range provide a high degree of flexibility. For example, the space planner can adopt a phased development process, initially building to accommodate the lower number, while retaining the capability to expand later to house the higher number (often by increasing the building's population density) if actual growth develops at the higher rate.

The methodology just described suffices for only the most general levels of facility planning. It can, for instance, be used to establish a rough estimate of the size of a new building that might be required to house future staff. The approach cannot, however, be used to identify precisely how large the building should be or how much space individual department occupants require.

Projection of Personnel by Comprehensive Projection Methodology

To assist departments in projecting future staff requirements, the following ten-step procedure is provided. It is based on procedures developed and used by the firm of Steinmann, Grayson, Smylie. The procedure provides a sequential analysis that enables groups with definable programs and services or quantifiable workloads to develop a projection of future personnel requirements. It also incorporates allowances for new programs and services, and it can be adjusted to accommodate changing policies, new legislation, and other changes.

The procedure is adaptable to different groups, upon making the appropriate changes

in terminology. Some groups have products, others have programs, others provide services, and still others simply have workloads to process. To that degree, terms such as *workload*, *program*, and *service* should be viewed as somewhat interchangeable. In some groups, for example, both workloads and service levels may serve as bases for projections.

STEP 1: Obtain mission parameters. In order to reduce the impact of subjectivity in projections, the process should begin with a statement of general activity, goals, and a likely range of organization-wide parameters (such as growth rates) from senior management. This permits development of a consistent enterprise framework for projections by department representatives.

STEP 2: Document department programs and functions. Each department should identify the variety of programs and services it provides. If a department is geographically separated, administrative (or headquarters) functions should be identified. The current number of personnel should be recorded, by specific program or function within each department. Organizational and functional relationships to other departments, sections, or divisions should be indicated.

STEP 3: Identify workload indicators. It is important to identify the significant workload indicators that might affect department staffing patterns. Appropriate workload indicators to quantify might include products shipped, permits issued, inspections conducted, service calls, client contacts, personnel supervised, applications reviewed, and forms processed.

STEP 4: Establish policies, goals, and objectives. Each department should document its goals and objectives for the coming years. This information should be reviewed by management for consistency with the organization's overall policy or strategic plan guidelines.

STEP 5: Identify variables affecting projections. It is important to identify variables, possible future changes, or assumptions that might significantly affect the projection of future workloads and related personnel levels. These assumptions, together with the probable effect of each on workload and personnel levels, should be documented.

Examples include macroeconomic and political variables, business changes, diversification, and competition. The individual effect of each variable or assumption can be tested against workload levels, after which new, adjusted projections can be developed. The cumulative effect of various combinations of assumptions can then be analyzed to determine the potential minimum and maximum impacts of each variable on workload projections.

It is beneficial for the department to identify which assumptions or combinations of variables are most likely to occur in the future. At the same time, it is helpful for the department to identify the groups of assumptions and variables that might sustain minimum and maximum projections of future requirements.

STEP 6: Quantify the impact of future service-level changes. The impact of changes in products, services, or programs provided—and the resulting workload projections—can be incorporated into the analysis at this point, and a forecast of future service levels and workloads can be developed. This quantitative workload for future years then serves as the basis for developing a projection of requirements to support that work effort in future years.

STEP 7: Perform a work productivity analysis. Having identified the work to be accomplished and the services to be provided, the department under study must determine how many employees are required for these purposes. A number of modifications to the employee productivity index may be valid. For example, more work can probably be accomplished by each individual in coming years as a result of automation or simplified procedures. Conversely, certain work efforts are becoming more complex and time-consuming. Environmental analyses and activity in the legal system are prime examples.

Any historical trend that indicates an increase or decrease in future productivity levels should be documented. To the degree possible, projections of future productivity levels—in terms of output or services provided per person—should be identified. A comparison of personnel levels and significant workload indicators at various times in the past should be made in order to identify relevant trends in productivity levels.

STEP 8: Perform an analysis of automation and efficiency opportunities. Many agencies or departments are likely to embark on programs to increase operational efficiency or productivity. Text processing, project management and scheduling techniques, computerization, new procedures, and new programs should be anticipated. To the degree that these concepts can be identified and anticipated to be employed during the coming years, their effect on reducing personnel requirements or increasing productivity per person should be incorporated into the analysis.

STEP 9: Develop a staffing requirements projection. Staff projections based on both future significant workloads and future productivity levels can now be made, after which these preliminary projections can be tested against current and historical data to determine their reasonableness.

STEP 10: Project administrative personnel requirements. Once total personnel levels for the department or agency are identified, the components of the total staff that are included in the administrative or headquarters function can be identified. Again, projections can be measured against current and historical data to test for overall relevance.

Projection of Space Requirements by Net Area Factor

The net area factor method provides a rapid but relatively inaccurate assessment of total space requirements. It results from multiplying typical average net area factors (a department's usable area, divided by the total number of personnel in that department) by total future staffing levels. The net area factor includes an appropriate allocation for special areas, unit equipment, common support areas, and circulation space. Total usable area requirements thus include all of the space within a building except the portions devoted to vertical circulation, building core and service elements, mechanical rooms, and significant structural or construction elements. Usually, the net usable area represents between 80 and 90 percent of the total gross area of a building.

An appropriate net area factor is multiplied by the projected personnel count. The personnel projection to be used can be derived by either the annual growth rate method or the comprehensive projection methodology. Multiplying a net area factor by the personnel projection derived from the latter methodology usually results in a more accurate and defensible estimate of space needs than does multiplying a net area factor by the personnel projection derived from annual growth rates. The annual growth rate method, however, is quicker and easier.

Following are rules of thumb for estimating general office space requirements. It is more difficult to generalize about industrial or manufacturing environments, due to their more specialized natures.

For estimating general office space requirements, a net area factor of between 170 and 190 net square feet (NSF) per person usually is appropriate. Area factors of between 170 and 190 NSF per person are quite reasonable in estimations of the total size of very large or-ganizations. On a more detailed level, however, it is possible to identify smaller departments that might have net area factors ranging from a low of 100 NSF per person to a high of perhaps 250 NSF per person.

Area factors include internal circulation within departmental boundaries. Included in the overall 170–190 NSF per person area factor are support spaces for conference rooms, filing and storage areas, libraries, and reproduction and photocopy areas. Usually, a significantly large special area that is not allocated to a particular department (for example, a major cafeteria, auditorium, garage, or public lobby) is not included in the net area factor and must be analyzed and added to overall space requirements individually.

Jim Steinmann suggests that a net area factor of between 160 and 180 NSF per person should be used for buildings where interiors are largely left to open-office planning, and where relatively extensive (10,000 square-foot and greater) floors exist. Net area factors of between 145 and 165 NSF per person are appropriate if widespread use of modular furniture systems is contemplated, and if no significant large special areas are included in the occupancy profile. Net area factors of between 175 and 200 NSF per person are appropriate if plans call for minimum efforts to be made to incorporate remodeling and to improve space utilization efficiency, or if the space is expected to contain a large proportion (in excess of 20 percent of the totals) of special areas and private offices. In general, larger overall space occupancies require a lower net area factor than smaller spaces do.

Constructing specific portions of the database can be simplified by the use of space takeoff sheets, representative results of which are shown in the output illustrated in figures 6-1 and 6-2. The sheet in the first figure is used to itemize personnel and space requirements; the sheet in the second figure is used to identify special areas and additional spatial considerations.

****

SWIMMER COLE MARTINEZ CURTIS
AREA NEEDS PROGRAM REPORT
DATE: 10/20/86   XYZ CORPORATION - LOS ANGELES HEADQUARTERS
                                                                                           PAGE   1

DEPT: CS05
      PRODUCTION CONTROL (VICKI COHEN)

POSITION / TITLE	LEVEL	W S CODE	SQ FT	YE86	YE87	YE88	YE89	YE90	CIRC FCTR	YE86	YE87	YE88	YE89	YE90
						QUANTITY						SQUARE FEET		
MANAGER		P03	144	1	1	1	1	1	1.23	177	177	177	177	177
SPECIALIST		P03	144	1	1	1	1	1	1.23	177	177	177	177	177
QUALITY ASSURANCE		OA1	152	1	1	1	1	1	1.23	187	187	187	187	187
SUPERVISOR		OA1	152	1	2	2	2	2	1.23	187	374	374	374	374
TECHNICIAN		OA3	71	1	2	3	3	3	1.23	87	175	262	262	262
ANALYST		OA3	71	6	6	7	8	8	1.23	524	524	611	699	699
SCHEDULER		OA3	71	1	1	1	1	1	1.23	87	87	87	87	87
OPERATOR		OA4	36	11	11	12	12	12	1.23	487	487	531	531	531
COORDINATOR		OA3	71	1	1	1	1	1	1.23	87	87	87	87	87

----- S U B T O T A L S -------------------

	24						2,000				
		26						2,275			
			29						2,493		
				30						2,581	
					30						2,581

SPECIAL AREAS ASSIGNABLE AREA (NEXT PAGE)   1,130   1,130   1,130   1,130   1,130

***** TOTAL ASSIGNABLE AREA ***********   3,130   3,405   3,623   3,711   3,711

AVERAGE ASSIGNABLE AREA PER PERSON   130   130   124   123   123

Figure 6-1. Database personnel projections. (Courtesy of Cole Martinez Curtis and Associates)

****

SWIMMER COLE MARTINEZ CURTIS
AREA NEEDS PROGRAM REPORT
DATE: 10/20/86   XYZ CORPORATION - LOS ANGELES HEADQUARTERS
                                                                                           PAGE   2

DEPT: CS05
      PRODUCTION CONTROL (VICKI COHEN)

SPECIAL AREA TITLE	AREA CODE	SQFT	YE86	YE87	YE88	YE89	YE90	CIRC FCTR	YE86	YE87	YE88	YE89	YE90
					QUANTITY						SQUARE FEET		
2DWR LEGAL FILE	SFVLG2	12	1	1	1	1	1	1.35	16	16	16	16	16
5DWR LEGAL FILE	SFVLG5	12	3	3	3	3	3	1.35	49	49	49	49	49
MICROFICHE CABINET		12	1	1	1	1	1	1.35	16	16	16	16	16
CARD CABINET		12	1	1	1	1	1	1.35	16	16	16	16	16
15X37 BOOK SHELVING		12	4	4	4	4	4	1.35	65	65	65	65	65
**** REMARKS: 42"H. 3 SHELVES													
15X39 BOOK SHELVING		13	1	1	1	1	1	1.35	18	18	18	18	18
**** REMARKS: 42"H. 3 SHELVES													
STORAGE ROOM		125	1	1	1	1	1	1.35	169	169	169	169	169
**** REMARKS: 27" DEEP OPEN SHELVING. 8 SHELVES HIGH.													
18X36 ENCL STRGE CAB	SSC1836	17	1	1	1	1	1	1.35	23	23	23	23	23
**** REMARKS: 42"H. 3 SHELVES.													
EQUIPMENT ROOM		100	1	1	1	1	1	1.35	135	135	135	135	135
**** REMARKS: FOR TAPE CLEANING EQUIPMENT													
30X60 TABLE		30	5	5	5	5	5	1.35	203	203	203	203	203
30X45 TABLE		23	5	5	5	5	5	1.35	155	155	155	155	155
30X60 DESK		30	4	4	4	4	4	1.35	162	162	162	162	162
36X60 DESK		33	1	1	1	1	1	1.35	45	45	45	45	45
26X35 TABLE		17	1	1	1	1	1	1.35	23	23	23	23	23
36X48 TABLE		26	1	1	1	1	1	1.35	35	35	35	35	35

SPECIAL AREAS ASSIGNABLE SQUARE FEET   1,130   1,130   1,130   1,130   1,130

Figure 6-2. Database special area projections. (Courtesy of Cole Martinez Curtis and Associates)

FACILITY MANAGEMENT SYSTEMS

Projection by Developing Space Allocation Standards

The preparation of a space requirements program is predicated on allocations of space to definable standard components. This approach gives a more accurate statement of requirements during the programming and planning phases of a project than does the net area factor method. Workstation space allocation standards provide the necessary basis for achieving this result.

Fairly general standards are required for developing a preliminary projection of future space requirements. More refined space allocation standards (providing specific information to the space planner) are required for developing detailed space plans and interior designs. This section discusses space standards appropriate for calculating detailed workstation area requirements.

The purpose of a space requirements program is to allocate space to people, furniture, and equipment. Standardization provides a consistent and necessary tool without which future planning is difficult; it also explicitly presents an opportunity to provide for the flexibility necessary to organizations that are continually changing. Standardization of requirements also suggests significant economies in purchasing workstation components and promotes space utilization efficiency. Finally, a comprehensive set of space standards saves time on space planning projects and relieves planners and designers from having to reinvent the wheel.

Workstation standards are developed by means of the following two-part procedure:

1. Job functions are analyzed to identify work tasks, equipment used and needed, performance levels, work methods, communication needs, and privacy requirements.
2. Each job classification is analyzed from the standpoint of its functional requirements. Specific attention is given to the type, amount, and size of each required element of the workstation. The components of the workplace are emphasized—not simply the amount of space in the workstation.

In developing workstation standards, emphasis is placed on identifying functional requirements, as opposed to satisfying requests from users. Functional requirements are the space or equipment needed to successfully complete a job function or task; requests for space or equipment, on the other hand, arise not so much from needs as from users' desires. Often it is not easy to distinguish requirements from requests, for the relationship between more or better lighting, space, furniture, and so on to increased productivity has yet to be solidly proven, despite much academic and institutional research and empirical study by both vendors and industry.

Quantification of functional requirements for typical workstations are developed in terms of the following components:

- *Work surface area:* primary, reference, and conference
- *Drawer storage:* box drawers, computer output storage, pencil, and other
- *Cabinet storage:* open, closed, and locked
- *Filing space:* legal, letter, drawing, and other
- *Bookshelves*
- *Primary and guest seating*
- *Special equipment:* word processing machines, teletypes, dictating units, calculators, microfiche readers, and so on
- Need for general *acoustical control,* plus individual visual and aural privacy
- Need for *security*

Workstation standards may vary from stand-alone desks to partitioned modular furniture in an open office setting to fully enclosed private offices. Examples of such standards are shown in figures C-11 and C-12 of the color insert. Although assignment of workstation standards usually is dictated by job classification, functional requirements are paramount. For example, two positions may share the same job title but have different duties, requiring different workstation components.

It is also important to keep standards to a minimum and not to invent too many standards.

This makes assignment of standards (as well as the ongoing process of managing them) simpler.

DATA INVENTORIES

Various items belong on a minimum listing of essential data inventories for space planning. These include record drawings and several types of supporting data.

Record Drawings

A fundamental element in the facility planning process is the assembly of record drawings, which serve to document both the present physical circumstances of a space user and the characteristics of the available space that the user is to occupy. Generally, three drawings fulfill the graphic requirements of record information: the record assignment plan, the as-built plan, and the core master plan.

The *record assignment plan* describes the spaces and their uses, as presently assigned. All space assigned to the specific user is graphically depicted in uniform scale and format, quantified in terms of net square-foot area and categorized by use and classification type. By reviewing the record assignment plan, the planner can determine existing patterns of use, calculate current ratios of space utilization and current area factors, and gain familiarity with the existing spatial quality of the user's facility. The assignment plan also provides documentation of departmental personnel and equipment allocations and relative area factors.

The *as-built plan* provides accurate definitions of the existing configuration and condition of space available for renovation. These drawings are self-descriptive and must accurately depict initial space configurations upon completion of the construction project. Field verification may be necessary to establish the accuracy and completeness of the as-built plan.

The *core master plan*, derived directly from the as-built plan after field verification, serves as the base plan from which all design and construction drawings originate. It must accurately depict all givens in the development of the new physical space solution.

Essential Supporting Data

In addition to record drawings of the types just named, data inventories must contain information on personnel, historical requirements, equipment, and future plans. Personnel data consist of the following types of information:

- Current personnel rosters
- Organization charts (both formal and informal), by name, level, task definition, organizational affiliation, and responsibility
- Functional description of the task responsibility of each group and subgroup

Historical data provide a context for current needs and suggest certain long-term trends. Such data typically include the following:

- Staff levels
- Space requirements
- Production quantities
- Task definitions

Equipment data consist of the following components:

- Inventories of furniture, fixtures, production equipment, manufacturing equipment, and special equipment
- Location of equipment
- Assignment of equipment
- Utilities
- Other special infrastructure requirements
- Maintenance, operating manuals, or catalogs

Future plan data help to identify the current thinking of top management about the organization's future direction. The following types of information are relevant:

- Future business plans and alternatives
- Projections of staff, equipment, furnishings, and space requirements
- Impact statements and/or sensitivity studies on the potential impact of new responsibilities,

changing workloads, and recent or pending legislation
- Procedural changes (resulting from known/expected reorganizations or policy changes)
- Use of automation and changing technologies

ALLOCATION

Space allocation involves assigning (matching) activities (people, equipment, tasks, meetings, classes, and so on) to zones (sites, buildings, floors, rooms, spaces, and so on). The goal of space allocation is to satisfy interaction requirements (material, communication, and work flow) while minimizing fixed costs (of moving, construction, and so on). Robin Liggett and William Mitchell (1981a, 1981c) have expressed the problem as being how to minimize a cost function consisting of fixed costs plus interactive costs. Their algorithm has been developed into a computer program called SABA (Stacking and Blocking Algorithms), produced by the Computer-Aided Design Group and licensed and utilized by several major CADD companies and computer vendors.

The product of an allocation is simply a definition of the space location or configuration occupied by each organizational unit represented in the space management database. Allocation represents both a technique (a process) and a product. The process of space allocation begins with an analysis of adjacency relationships among the various organizational units, and it concludes with products: a matrix, a space allocation plan, and specific assignments of space for the future.

The subsections that follow discuss techniques that can be employed by the space management organization in analyzing adjacency relationships on a large scale (where different communities or buildings are to be considered) and on smaller scales (within buildings, on floors, and among individuals). These techniques represent a management resource distinct from the (mostly computer-based) tools discussed elsewhere in this book.

Locational Strategies

Locational strategies are long-term plans for optimal placement of departments and agencies. In the absence of such planning, space users tend to be placed haphazardly, either where space is currently available or where it can most easily be made available through construction, renovation, or lease. The results often are inefficient and uneconomical operations, inaccessibility, and the need for more frequent departmental relocations.

Macrolocational Analysis

At the broadest level, locational analysis is concerned with which community (or country) is the most appropriate locale for an organization. The space management organization is often expected to advise senior management on major relocation concerns such as where a new operations center, manufacturing or distribution activity, or branch office should be located, or how to resolve policy issues related to centralization or decentralization.

Generally, when a question of locale selection is addressed, the following issues should be considered:

- Availability of a qualified labor pool
- Labor rates
- Quality of labor
- Housing cost
- Housing availability
- Willingness of key management personnel to relocate
- Willingness of employees to relocate
- Availability of raw materials and supplies
- Relative proximity to other operations of the organization
- Relative proximity to required distribution centers
- Cost of operations
- Cost of construction
- Community quality as an incentive for employee retention

Major relocations require thorough analyses of associated life-cycle costs. The impact of interest rates, operating and housing costs, and other factors may be substantial. The ability to acquire key skills, staff, and management personnel in a new community or in a different location within a city must be thoroughly considered. A 1982 study of large government and corporate organizations by Jim Steinmann indicated that relocation costs typically exceed $40,000 per employee as a reasonable rule of thumb.

A significant body of literature is available to facility managers on the subjects of researching and assessing site and community selection. Publications of the Industrial Development Research Council (IDRC), a professional organization, offer an excellent introduction to both subjects.

Site Selection

After a general location is specified, the next step is to undertake a more detailed search for a particular site. Although the site selection study must focus on relatively immediate and detailed issues, the general topics raised are similar to those listed above: access; proximity to housing, labor, and markets; cost of land acquisition, development, and leasing; general growth patterns; flexibility to support future expansion; and so on.

Computer-assisted site selection based on demographic characteristics recently has become commercially available and is likely to become more prevalent as time goes on. Another example of computer-assisted site selection (within a large site) is *overlay mapping*—a technique in which many variables are overlaid (often graphically) in logical combinations to identify optimum places for construction.

Again, a detailed analysis of the alternatives (including a careful quantification of the costs involved in alternative locations) must be developed by the facility management organization, in order for it to advise management as beneficially as possible.

Stacking and Blocking

Once the general location for a facility is decided and the identity of the organizational units that will occupy that location and/or site is determined, a microanalysis of adjacency relationships for departments and work groups is needed.

As more detailed space plans are developed, the location of each department within each building is defined through the creation of *stacking plans* (sometimes also called *stacking profiles*) and *block plans* (sometimes also called *area blockouts* or *blocking plans*). Stacking plans are vertical sections (profiles or slices, taken vertically) through each building, showing all floors and identifying the occupants by department and area or work unit on each floor. Block plans are single-line drawings or sketches of the horizontal (plan view or floor plan) of each floor, illustrating the approximate location of departments on each floor and the estimated space required by each department.

Sometimes block plans are preceded or even replaced by bubble diagrams (rough sketches showing relative size and adjacency but not physical form) or adjacency circle diagrams. Sometimes adjacencies between departments are shown as lines in which the intensity of the adjacency need is represented by the thickness of the line. These *adjacency line diagrams* sometimes are overlayed onto block plans. Examples are shown in figures C-13 through C-15 of the color insert.

Stacking and block plans often are prepared after the assignment of departments to buildings has already been approved. Because numerous factors are involved in the final placement of departments (including actual levels of required interface/adjacency, departmental expansion requirements, the timing of space availability, and the desire to maximize space utilization), stacking plans and block plans often are created through the use of computer models, as discussed earlier. Figures C-16 through C-19 of the color insert show examples of such computer-generated plans.

Area assignment diagrams (yet another term for both stacking plans and block plans) can be employed to confirm that building assignments made during master planning phases of study are indeed feasible within a particular structure or structures.

Goals of Locational Analysis

The primary goal of locational analysis is to identify the adjacency links crucial to the productivity of the organization. Although it is important to place departments engaged in regular and significant interchanges of information in close proximity, the following issues should also be considered:

- The desire to consolidate currently fragmented departments
- The desire to locate many (if not most) departments in a central building or complex
- Total current available space, by geographic area
- The cost of leases and options required to satisfy certain adjacencies
- The remodeling and relocation costs related to satisfying adjacency requirements
- Requirements for visitors and public access
- The desires of management

The desires mentioned in the final item above often are a result of organizational style and culture. For example, one corporate president might want all of the organization's departmental heads centralized ("at headquarters"); meanwhile, another president might want each department head to stay with the rest of his or her department (decentralized).

In some cases, departments may be placed in less-than-optimal locations. This sometimes is done for such simple reasons as that a particular space is already owned or leased, for which no better economic use exists.

Economic issues are highly relevant in deciding the placement of special-use and shared-use facilities. For example, a first-floor cen-

tralized location might be considered for a computer room because of its heavy interface with a number of departments, or because infrastructure services (power, air conditioning, delivery access, and so on) are more readily or economically available on that particular floor or some portion of it. On the other hand, security considerations, plus requirements for raised floors, spatial clearances, and special utility provisions, may suggest an alternative placement. Indeed, major computer facilities are rarely relocated merely to improve functional adjacencies. Other examples of special-use facilities include auditoriums and two-story or high-bay spaces such as warehouses. As a result of structural considerations, a decision may be made not to locate these types of facilities within or adjacent to general office space, even though functional adjacencies may so suggest. If departments can share a special-use facility such as a library or training room, the organization may find it advantageous to locate these departments in adjacent space for this reason, even though the departments have no other functional adjacency requirements.

The computer algorithms available to address these sorts of problems do so in a variety of ways. SABA, for example, represents some of these issues in the form of fixed costs.

An optimizing algorithm such as SABA can (theoretically) produce a near-perfect response—whether or not this approaches being a near-perfect solution. The algorithm can demonstrably come close to optimizing the response to input, but the degree to which the input is accurate, complete, and relevant to the problem in need of solving is the responsibility of the user. Only if the input data reflect all known criteria will the result be good. The challenge to the facility manager is to formulate the input to the computer in such a way that all important considerations are included, and the considerations are weighted properly in relation to each other. The latter task is easy to say but extremely difficult to do. The proper weighting of input variables to spatial assignment algorithms is every bit as important to the quality of the solution as the power and accuracy of the algorithm itself.

At present, few computer applications attempt to develop an optimized stacking plan or block plan automatically. Most allow the user to make the assignments, produce an alternative arrangement, and—by manual inspection or computer-assisted weighting and evaluation—determine the best fit. This process uses the computer as a visualization tool, and sometimes as a very accurate scorekeeper. A few computerized programs, however, are designed to optimize adjacency relationships among a large number of organizational units in a large multilevel building or buildings. Similar application programs have been developed to assist industrial engineers in allocating space in an industrial layout or assembly-line facility. These applications are extremely beneficial when large (in excess of 40,000-square-foot) floors are under planning.

Methodology to Evaluate Locational Requirements

Locational requirements should be developed in the course of a careful examination of relationships among personnel, equipment, and space. An adjacency relationship should be established on the basis of the expected future levels of interaction, work flow, and contact of organizational units. Future levels of telephone and mail activity probably should not be considered.

Documenting the current degree of interaction, face-to-face contact for meetings or discussions, and so on, may not reflect future environments. For example, the current proximity of one organizational unit to another may have caused a high degree of interaction between the two. In the future, a different series of group locations may suggest that telephone, electronic mail, voice mail, video conferencing, computer interfaces, or some other form of communication may suffice to satisfy adjacency relationships. Conversely, if two interacting departments are located relatively far apart, they may not have much day-to-day contact, even though the business transactions and relationships between them are important and might be improved if they enjoyed closer proximity.

Thus, survey vehicles (questionnaires or other data-gathering means) employed by the facility management organization in projecting future interaction levels for establishing adjacency relationships are crucial. These data should document current adjacency relationships, but more important they should attempt to project need (what future level of interaction will be required) over the planned life of the occupancy.

Technologies such as electronic mail, voice mail, teleconferencing, and video are beginning to have a major impact in the business environment. Although it is too early to predict the outcome with precision, these technologies seem likely to reduce significantly the need for physical adjacency in a large number of working groups and working environments. In any case, perhaps ten to twenty years will pass before a significant fraction (up to half) of the group adjacency needs of "knowledge workers" is met by technology in lieu of physical proximity.

More Detailed Planning

Once block plans have been determined, locational analysis continues at a more detailed level (room to room, or one individual or piece of equipment to another). These microarea analyses can be aided by recourse to a computer system for optimizing the location of departments and for assisting in planning the detailed layout and placement of workstations and pieces of equipment. There are a variety of very good computer-aided design and drafting (CADD) systems on the market to serve this purpose, as well as to produce contract documents (plans, elevations, schedules, specifications, and engineering drawings).

Checklist of Master-Plan Procedures

Major considerations to be evaluated in developing a master plan include the following:

- *Operations:* the day-to-day functioning of departments and the interface between them
- *Public accessibility:* the degree to which a department is visited by the general public
- *Common clientele:* the number of visitors who need to see more than one department during a particular visit

As part of the process of assembling a database for identifying adjacency requirements, each department or agency responds to questions designed to identify why each should be located in a particular place (as opposed to an alternative location). The following questions, which have been generalized from questions appearing on several commercial space planning firms' questionnaires, may be included:

- List all departments that are visited more than five times per week by your department, and estimate the actual number of trips.
- List all departments that visit your organization, and estimate the number of weekly visits.
- Identify all departments that ideally should be located in the same building as yours.
- Identify all departments that ideally should be located in the same complex as yours.
- To what extent does your department need to be located in or near a central building or complex?
- Identify the number of daily public visitors to your department, the average duration per visit, and the general purposes of the visits.

Other questions can be considered through a combination of questionnaire input and management policy as determined at interviews. The following examples fall into this category:

- Do functional requirements dictate that face-to-face contact is essential, or would messenger service suffice? (Interaction patterns often result from current departmental proximities and/or some employees' desires to leave their workplace occasionally for a break.)
- What proportion of a department's employees actually make interdepartmental trips? (Often only

one or two individuals make most such trips, and their situation alone would not warrant relocation of an entire department.)

- What are the organizational levels of the most active visitors? (Obviously, labor rates for high-level employees make their travel time more expensive than that of clerical-level staff. In addition, higher-level staff usually have more crowded schedules, and many regularly require quick resolution of problems through face-to-face interaction.)
- Is the flow of documents or material between departments large or small?
- Would locating two departments together hinder operations by increasing the likelihood of unnecessary visits? (The personnel department, for example, may prefer to remain geographically remote for this reason.)
- Are interaction patterns seasonal, or do they remain constant throughout the year?
- Do parking, public accessibility, and/or image-projection considerations outweigh other adjacency criteria?
- Do departments work together in a regular or sequential process?
- Would relocation cause so much temporary disruption of normal work activity or be so costly as to cause the disadvantages to outweigh the advantages?
- Do any departments have similar space requirements or utilize the same records? (If so, it may be appropriate to colocate these departments— for example, warehousing and records storage activities that require high bay space. Departments may also be able to share auditoriums, large conference rooms, computer facilities, and other space and equipment. In general, the organization should attempt to locate departments together if such placement would facilitate the shared use of high-cost special areas.)

Responses to these questions are tabulated, and interaction levels and preferences are identified in rank order for all departments.

To identify adjacency priorities from the mass of raw data gathered, an interaction matrix is generated that displays interfaces between departments. Figures C-20 through C-22 of the color insert show examples of interaction matrices. Once the ideal relationships are identified, a system designed to prioritize adjacencies is brought into play. In most cases, obvious physical constraints dictate the amount of desired adjacency that can be achieved. Therefore, prioritization (or weighting) of adjacency needs is required.

For example, a department's need to locate at a particular site because of shortened travel time to other departments may be rendered secondary to the need of other departments to be at the same site for operational or functional reasons (such as material flow). Priorities may reflect criteria that promote cost-effective operations, convenience, aesthetic values, or some other objective established within the organization's administration; as such, they should be carefully reviewed as a matter of policy and procedure prior to use.

Ongoing Planning

In practice, master planning is seldom done as a single, comprehensive, integrated process. Opportunities for optimizing all adjacency and work relationships by means of a single clean-slate master plan generally do not arise more than once (if at all) in a facility manager's career. The opportunity to oversee a total organizational move into new quarters is rare indeed. Much more often, locational and adjacency analysis is a matter of working within well-established constraints and settled conditions of existing departments and equipment—a task that requires moving the chess pieces slowly, one at a time.

The gradual nature of this process does not invalidate locational analysis, master planning, or the use of optimization algorithms and models. To the contrary, an analysis of the organization's current status and future goals is essential if the facility manager is to ensure that each new move and construction project will improve the overall situation. Indeed, computer models may be most effective and helpful in these overconstrained situations, where the question "what is the optimum location for each of our departments?" is relatively moot and where the more important and practical question is "what is the least-cost path from our present set of locational assignments?" Computer models only now are beginning to be commonly used to address these more complex problems.

Update

Information in the database should be updated periodically according to whatever update schedule (annual updating is often used) is established for revising information in the database to include material gathered in the same process or at the same time. If a space user requests significantly more resources than have been planned for, however, or if a new location is requested, an update of the department involved may be necessary. The space user should then be required to complete a survey form or similar vehicle to answer the previously mentioned adjacency-related questions.

The update procedure should contain the following seven major steps:

1. Gather raw data as required.
2. Identify fixed departmental locations.
3. Identify and prioritize the actual interface requirements of the remaining departments and of the fixed departments.
4. Identify large and growing departments, and specify their expansion potential.
5. Locate large and growing departments on the basis of priorities identified in Step 3 above, taking into account expansion potential and other considerations mentioned throughout this chapter.
6. Locate relatively autonomous, small, stable departments in spaces between large and growing departments, in order to satisfy the potential expansion requirements of the latter by holding space reserves for them.
7. Provide for relocation of the small departments as required.

Good automated allocation algorithms should take much of the preceding systematic approach into consideration.

Applications

The technique of adjacency analysis (or on a larger scale, locational analysis) can be applied to the following endeavors:

• Determining which community to place an organizational unit in
• Carrying out the site selection study process
• Analyzing alternative lease spaces for occupancy
• Determining which organizational unit(s) to place into each of various buildings
• Determining which organizational units are on the various floors of a building in a multilevel application
• Determining the allocation of space on a particular floor (area assignment plans)
• Developing detailed space plans

Probably the most important applications of adjacency or locational analysis (and the most influential level of its product) are the following, performed when new construction is contemplated:

• Determining whether new construction is needed
• Determining the optimum location (country, city, campus, site, and so on) for the new construction
• Determining the size and configuration of a building, if new construction is contemplated
• Determining the optimum floor size in a building that is to be designed or modified

Of less importance is the use of a formalized approach in determining the locational arrangement of personnel within a smaller (less than 5,000 or 10,000-square-foot) organizational unit. Gathering information that is too specific and detailed about the level and frequency of interaction from one individual to another is not a particularly productive or useful activity when the task is to determine how to allocate space in a typical (smaller) highrise floor. This is because small floors (which account for probably 90 percent or more of all space occupied by large organizational units in new buildings) rarely involve relative adjacency relationships as a primary constraint.

Examples of more pressing factors in these instances include the location of a central core, the core-to-window-wall distance, the planning module, required circulation and fire-exiting patterns, specific constraints on space assignments (such as executives in the corner offices, private offices on the perimeter, or computer and support facilities near the service elevator), and so on. These constraints, which are often identified early in the study, normally provide the space planner with sufficient information to develop area-assignment block plans for the various suborganizational units occupying a floor—and to do so in a very rapid manual or CADD fashion. Rarely does quantification of the relative importance of adjacency relationships among individuals or small subgroups on a floor lead to development of a more appropriate area-assignment block plan.

Many current computer applications rely on this type of microlevel adjacency analysis, but the resulting improvement in the quality of the projections rarely justifies the time or expense of developing it. On the other hand, very good use can be made of the workstation design information often gathered for use by these computer programs. Information can and should be gathered relating to individuals' work styles, habits, equipment needs, and so on, in order to assign each individual to the appropriate workplace, and to give each individual the proper furniture, equipment, and environment.

COMPUTER PROGRAMS

Early Programs

A sample of computer programs that have been developed for spatial allocation problem solving (in rough chronological order of their introduction) might include CRAFT, ALDEP, DOMINO, and STACKing Plan. A brief discussion of each of these programs follows. William Mitchell provides an excellent and more extensive discussion of these and other *automated spatial synthesis* algorithms in *Computer-Aided Architectural Design*. These early programs are rarely used commercially anymore, but they form the basis for virtually all of the techniques that are used.

CRAFT: This program, whose acronym stands for Computerized Relative Allocation of Facilities Technique, requires the user to input an initial layout within a defined building boundary; then it follows a step-by-step strategy of evaluating small prospective modifications to the layout, with the goal of decreasing the overall circulation cost. CRAFT, developed in 1963 by G. C. Armour and E. S. Buffa, was most appropriate where optimization of circulation constituted the dominant layout consideration. CRAFT-like programs often lead spaces to be twisted into unacceptable shapes, and they tend to be slow and expensive in execution when used for realistically large plans. CRAFT is, however, generally acknowledged as the first practical computer program for automated spatial synthesis.

ALDEP: Another early program, Automated Layout Design Program was developed by J. M. Seehof and W. O. Evans in 1967. It is an interesting example of the use of a sampling technique. ALDEP follows a strategy of rapidly generating random alternative layouts, which are evaluated against proximity requirements between spaces. Layouts that satisfy certain threshold standards are printed out. This is fast, cheap, and produces numerous alternatives for consideration. Unfortunately, it (like CRAFT) suffers from a lack of shape constraints.

DOMINO: This program was developed by William Mitchell at UCLA. It locates one module at a time in a grid, according to heuristics (rules of thumb) derived from designers' experience, in order to generate layouts that comply closely with area, shape, and adjacency requirements. The process is relatively fast and cheap, and it produces many possibilities of satisfactory quality.

STACKing Plan: This program was developed by Charles Reeder at UCLA and later further developed for Morganelli-Heumann & Associates. Like DOMINO, STACKing Plan was designed to utilize heuristics. While DOMINO's heuristics were applied to the two-dimensional problems of single-floor plans, STACKing Plan's heuristics were applied to the one-dimensional switching problems in vertical stacking of multifloor buildings. STACKing Plan is still being used by CHA Interior Architects, a successor of the firm for which the program was first developed.

Current, Nonoptimizing Algorithms

Figures C-23 and C-24 of the color insert depict block plans produced by representative present-day commercial implementations of heuristic and other nonoptimizing algorithms.

Current Optimizing Algorithms

Robin Liggett and William Mitchell's optimizing algorithm, embodied in the Computer-Aided Design Group's SABA, is used by several major CADD vendors and is something of an industry standard methodology. An advanced mathematical concept called *implicit enumeration* is used by SABA to produce mathematically provable near-optimum results in acceptable amounts of computer time. Examples of results generated from this algorithm are shown in figures C-25 through C-27 of the color insert.

Stacking and Blocking in Sequence

Initially, stacking is performed before blocking is. Often, however, insights gained from viewing the block plans are used to modify the input criteria for revised stacking. Figure C-28 of the color insert shows the normal manual sequence for proceeding from stacking plans to detailed space plans.

The introduction of automation has made it possible to perform site allocation, stacking, and blocking simultaneously. Figure C-29 of the color insert illustrates output from SABA to be used in solving a stacking plan and block plans in this manner.

AUTOMATION OF FURNITURE AND EQUIPMENT SELECTION AND DETAILED LAYOUT

The selection of furniture and equipment is normally performed by an interior designer who must rely heavily on prior experience and rules of thumb. Once certain key decisions about standards and overall feeling and quality are made, detailed selections and specifications often follow fairly mechanically. In principle, great potential exists for automating this routine, detailed decision-making, using such standard data-processing resources as decision tables. For the most part, computer-aided systems currently used to automate this process exploit the storage and retrieval capabilities of the computer. For example, furniture systems from competing manufacturers are sorted and listed by modularity, finishes, and other attributes.

An interesting standard is developing in the computer-aided specification of contract furniture. Computer-Aided Planning (CAP), a Grand Rapids, Michigan, firm, has developed a simple, keyword-driven method of storing and specifying systems furniture. Known as CAPSIF (for Computer-Aided Planning Standard Interchange Format), it enjoys use by most of the major manufacturers of contract systems furniture.

Detailed layout of offices and workstations involves the precise placement of furniture and equipment in response to numerous constraints. This is normally a task performed by experienced space planners and drafters, but partial automation is now becoming possible. Most of the major vendors of computer-aided design and drafting (CADD) systems have begun to field systems that address these matters.

COMPUTER-AIDED DESIGN AND DRAFTING (CADD) SYSTEMS

Facility management functions generally are concerned with the nature and contents of architectural space. The principal source of information is usually as-built drawings (drawings depicting the spaces as they actually were constructed—as opposed to how they were designed to be constructed—and modified, often drawn long after construction is complete), plus information allied to particular aspects (attributes) of the drawings. The drawings are invariably architectural plans, sections, and elevations of the facilities. Allied information (net, gross, usable, assignable, and rentable areas; comparable data on occupants; and so on) is usually maintained in the form of lists of information kept in filing cabinets or as annotations to the drawings.

Maintaining a system that relies on two principal sources of information has several clear disadvantages:

1. Having two information sources—one of which is derived from the other—raises the necessity that concurrency of the data be maintained. This means that any change in one source must be reflected in the other, at the expense of considerable extra time and the hazard of additional errors from inconsistency.
2. Significant parts of the attribute information are present on the drawings and properly must be derived directly from the drawings by (often manually) measuring, calculating, or counting. This procedure, too, is extremely time-consuming and susceptible to error.
3. Attribute information, as traditionally maintained, usually suffers from the trait of information loss. For example, maintaining a record of the rentable area on a floor as a number in a list does not retain facts about the shape of the area, access to the area, the area's position, or its adjacency to other areas. These are all important related items of information.

Thus, to connect these two sources of information together—as well as to access and present them in a manner suited to what they actually represent (part of architectural space)—would be highly beneficial. More precisely, items of facility information should be found and presented by reference to pictorial information (such as an architectural drawing).

CADD systems use architectural space as their information base. In present commercial CADD systems, information generally is organized by location. Although primarily intended for the production of working documents concerned with the design of architectural space, some of the more sophisticated CADD systems available today include features well-suited for facility management. The reasons for acquiring these systems in an architect's office are at least partly applicable to a facility management situation.

The practical and economic justifications for acquiring CADD systems for architectural work (the production and modification of contract documents, that is, drawings and specifications) are well established. Here we are also discussing the additional possible applicabilities for ongoing facility management.

The recent surge in acquisitions of CADD systems by architects has been fueled by their need to increase productivity. Essentially, productivity benefits are achieved by replacing labor (professional, technical, and clerical staff) with capital (computer hardware, software, and data) in the following types of tasks:

• Producing new data, such as design decisions and results of analyses
• Retrieving historical data, such as drawings from previous projects
• Formatting and reporting data to generate various types of drawings and printed reports, such as counts and takeoffs
• Correcting and editing data in drawings, reports, and databases
• Transferring data from one location to another, such as from the design office to the construction site

Sophisticated CADD systems can maximize productivity in the tasks just named, if they can provide the following services:

• Productivity assistance in automating the drafting of architectural and engineering drawings (not a primary function of facilities management, but nonetheless of importance to a facilities manager)
• Data takeoffs and calculations for reporting and querying purposes, and generation of printed reports, schedules, specifications, and drawings (rather than manual counting, typing, and cut-and-paste editing)
• Checking of design proposals against externally specified criteria (such as a space program)
• Sorting and tabulation of data for different reporting purposes (for example, sorting a furniture schedule by location, type of item, time frame for ordering, or lease expirations by location)

Besides offering the productivity enhancements obtainable through the preceding automated processes, a CADD system can provide other important benefits. First, the information base is kept in one place and is readily accessible (within desired security provisions) to all who require information. Type of access (for example, read versus write) can be controlled, and drawings maintained on a computer system do not deteriorate with time or take up as much floor space in storage. Second, a generalized drawing system is capable of producing a wide range of graphical representations of numerical data. The rapid growth of business graphics is a response to management's need to have information presented with clarity and precision. Histograms, pie charts, bar graphs, and so on, can be produced to represent the numerical information present in a facility management system. Plots of future growth of floor area requirements versus actual space available, comparisons between organizational groupings, rent income projections, and other representations are possible with this feature. Figures C-30 through C-34 of the color insert show examples of these sorts of reports.

CADD systems are still in their adolescence and may not as yet be clearly or generally cost justified in facility management contexts. Their increasing use (and cost-effectiveness) over time, however, is inevitable. Consequently, facility

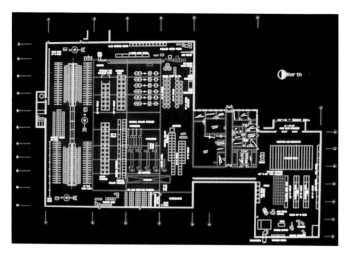

Figure C-1. CADD output.
(Courtesy of Auto-trol Technology)

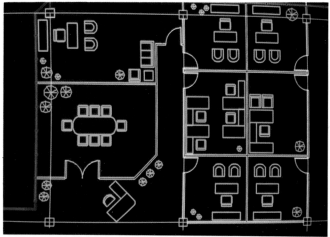

Figure C-2. CADD output. (Courtesy of Sigma Design)

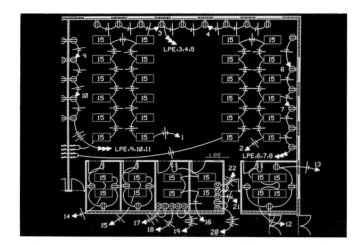

Figure C-3. CADD output. (Courtesy of Intergraph Corp.)

Figure C-4. CADD output.
(Courtesy of California Computer Products, Inc.)

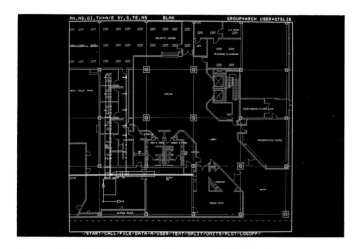

Figure C-5. CADD output. (Courtesy of CADAM, Inc.)

Figure C-6. CADD output. (Courtesy of General Electric Calma Co.)

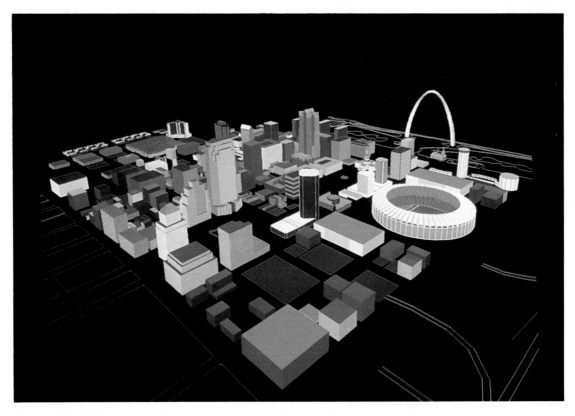

Figure C-7. CADD output. (Courtesy of Hellmuth, Obata & Kassabaum, Inc.)

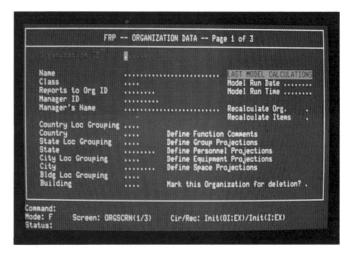

Figure C-8. Screen forms. (Courtesy of CADAM, Inc.)

Figure C-9. Screen forms.
(Courtesy of Computer-Aided Design Group)

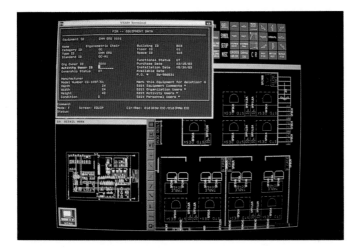

Figure C-10. Screen windows: forms and CADD.
(Courtesy of Auto-trol Technology)

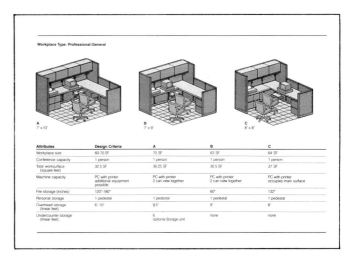

Figure C-11. Workstation standards.
(Courtesy of Gensler & Associates)

Figure C-12. Workstation standards.
(Courtesy of Hellmuth, Obata & Kassabaum, Inc.)

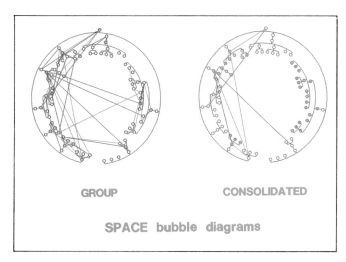

Figure C-13. Adjacency circle diagrams.
(Courtesy of Hellmuth, Obata & Kassabaum, Inc.)

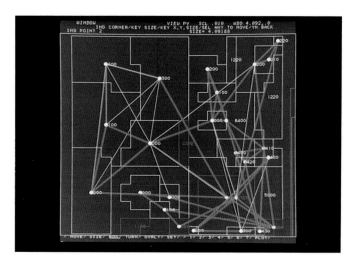

Figure C-14. Adjacency line diagram and plan.
(Courtesy of CADAM, Inc.)

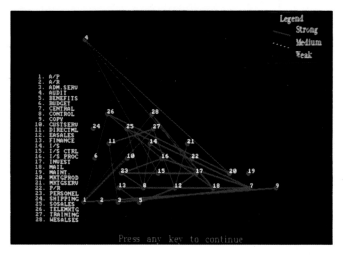

Figure C-15. Adjacency line diagram.
(Courtesy of Micro-Vector, Inc.)

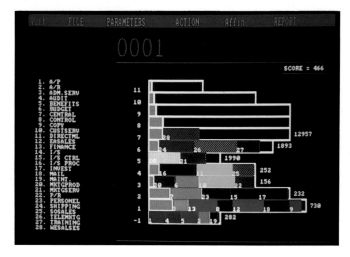

Figure C-16. Stacking plan.
(Courtesy of Micro-Vector, Inc.)

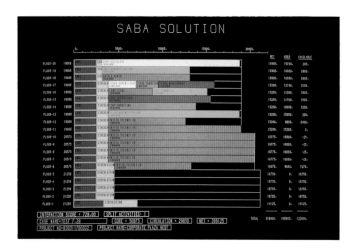

Figure C-17. Stacking plan.
(Courtesy of California Computer Products, Inc.)

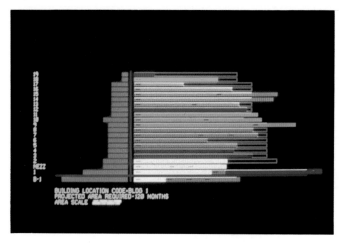

Figure C-18. Stacking plan.
(Courtesy of Intergraph Corp.)

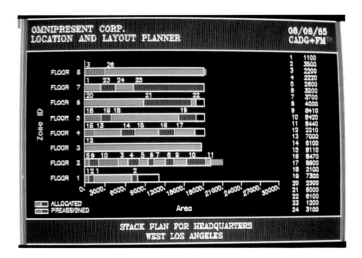

Figure C-19. Stacking plan.
(Courtesy of Computer-Aided Design Group)

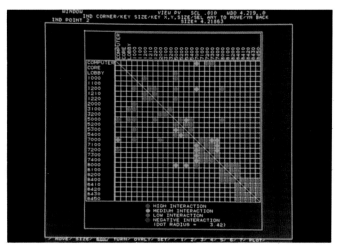

Figure C-20. Interaction matrix.
(Courtesy of CADAM, Inc.)

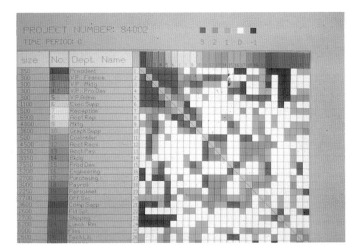

Figure C-21. Interaction matrix.
(Courtesy of Sigma Design)

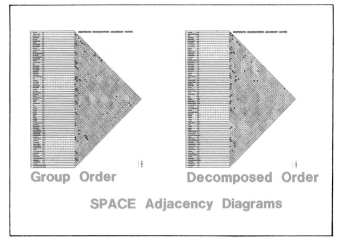

Figure C-22. Interaction matrix.
(Courtesy of Hellmuth, Obata & Kassabaum, Inc.)

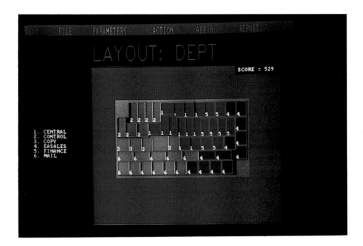

Figure C-23. Block plan.
(Courtesy of Micro-Vector, Inc.)

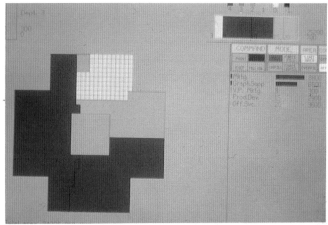

Figure C-24. Block plan. (Courtesy of Sigma Design)

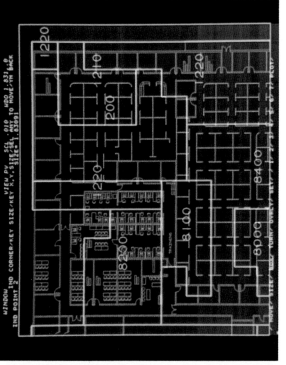

Figure C-26. SABA block plan. (Courtesy of CADAM, Inc.)

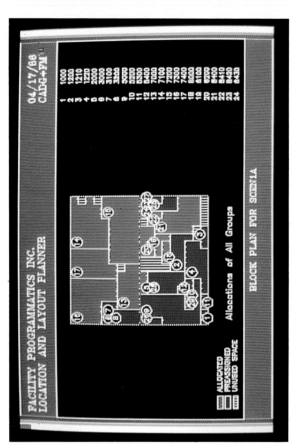

Figure C-25. SABA block plan.
(Courtesy of Computer-Aided Design Group)

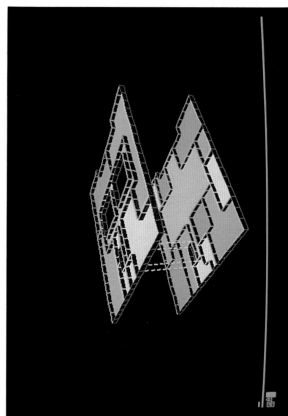

Figure C-29. Simultaneous stack and block concept.
(Courtesy of Computer-Aided Design Group)

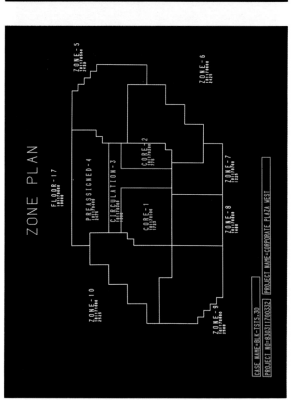

Figure C-27. SABA block plan.
(Courtesy of California Computer Products, Inc.)

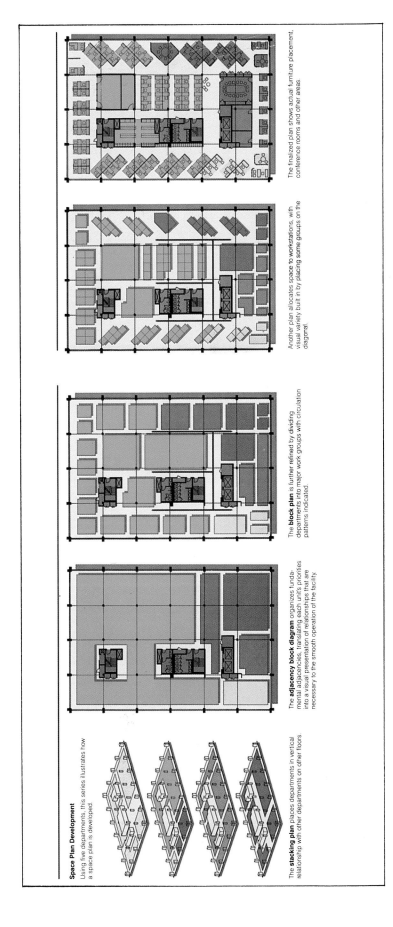

Space Plan Development
Using five departments, this series illustrates how a space plan is developed.

The **stacking plan** places departments in vertical relationship with other departments on other floors.

The **adjacency block diagram** organizes fundamental adjacencies, translating each unit's priorities into a visual presentation of relationships that are necessary to the smooth operation of the facility.

The **block plan** is further refined by dividing departments into major work groups with circulation patterns indicated.

Another plan allocates space to workstations, with visual variety built in by placing some groups on the diagonal.

The finalized plan shows actual furniture placement, conference rooms and other areas.

Figure C-28. Space plan development. (Courtesy of Gensler & Associates)

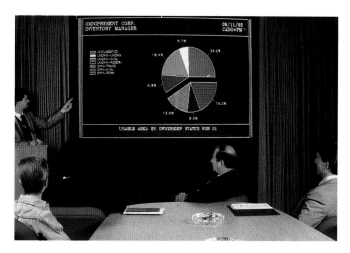

Figure C-30. Usable area, by occupancy status.
(Courtesy of CADAM, Inc.)

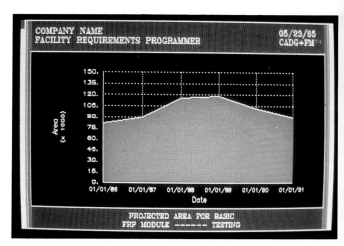

Figure C-31. Projected area requirements.
(Courtesy of Computer-Aided Design Group)

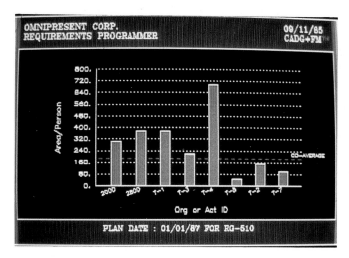

Figure C-32. Area per person.
(Courtesy of Computer-Aided Design Group)

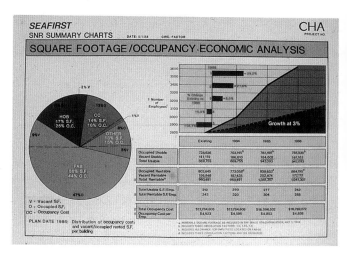

Figure C-33. Economic analysis of square-footage occupancy.
(Courtesy of CHA)

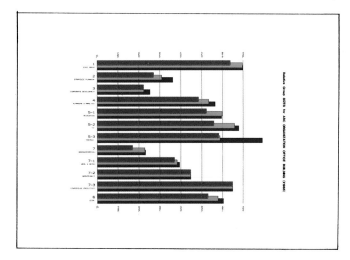

Figure C-34. Relative group sizes.
(Courtesy of Hellmuth, Obata & Kassabaum, Inc.)

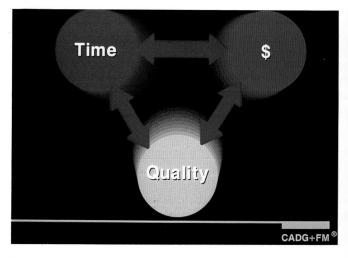

Figure C-35. Cost, time, and quality.
(Courtesy of Computer-Aided Design Group)

management systems must be designed as decision-support systems, in anticipation of eventual close coupling with CADD systems. Whether a CADD system is part of a facility management system or not (and a facility management system can function quite adequately and cost-effectively without one), eventual integration with a CADD system is almost inevitable. The benefits obtained from a fully computerized facility management system with or without CADD can be summarized under the following general areas:

- *Speed:* Access to and modification of data are achieved at very high speed. Access to, updating of, and reporting of information on modern output devices are orders of magnitude faster than manual methods.
- *Accuracy:* The accuracy of accessed and computed information cannot be equaled by manual methods.
- *Consistency:* A computer performs calculations without regard to time or place. With proper system design and programming, data can be stored and accessed absolutely identically, no matter where or how many times these processes occur.

Present commercial CADD systems are excellent, efficient, and effective decision-execution and decision-implementation tools. A good facility management system should be a decision-support tool. While decision support and decision execution certainly could be commercially packaged (and quite a few good reasons exist for doing so), this is only now just beginning to occur.

The other side of this coin is that any computer system—no matter how comprehensive and sophisticated—will fail dramatically (and expensively) if its limitations and capabilities are not understood. A computer cannot process or report information it does not have; and it will unhesitatingly spew out endless amounts of out-of-date information, without caveat. When computer systems are established, institutions often have to be reorganized to accommodate the new information base. The internal organization and the nature of work performed by a particular institution sometimes

preclude the optimum use of computer systems. It is worthwhile to repeat here that good procedures (and not necessarily good automation) define good facility management systems.

CONTRACT DOCUMENTS

Documentation of space planning schemes normally takes two forms: drawings (sometimes simply called plans), which show locations and arrangements of workstations, spaces, furniture, and equipment; and specifications, which record construction, selection, and assignment decisions. Computers can generate both forms of documentation at high speed and low cost.

Drawings

Automated drafting technology is highly developed and is widely employed in many fields of architecture, engineering, and interior design. Applying this technology to the production of layout drawings involves no technical difficulty; suitable hardware and software are commercially available in the form of many CADD systems.

The economics of automated drafting for this type of activity, however, requires careful analysis. Costs of data entry and equipment amortization can easily outweigh the potential benefits to be gained, as many organizations have discovered to their chagrin. In the past, automated drafting was at best a marginally cost-effective proposition for all but the largest organizations. The rapidly decreasing costs of computer hardware, the steadily increasing sophistication of CADD software, interfaces, and data-entry methods, and the rising cost of skilled labor, however, have shifted the balance. In many contexts, even relatively small organizations may find that automated or computer-assisted drafting is now feasible.

Specifications

Specifications can be produced by means of the report-generating facilities of database management systems and of a few CADD systems; eliminating hand typing in this way can cut days from the completion time of a project. Sorting and formatting facilities allow a wide variety of reporting, specification, and ordering documents to be generated economically from the same database. These systems also allow cost projections and analyses to be made with relatively little effort. In addition, the computer can conserve the specification writer's time by assembling packages of text into semifinished form for final editing and approval—often in response to a short, building-type questionnaire.

POSTOCCUPANCY ACTIVITIES

Following building occupancy, the space, furniture, and equipment databases generated during the design process may be used as the foundation of the organization's ongoing facility management system. Clients already have begun to demand this information (in machine-readable form) from their architects, and this trend will continue and expand. The growth of clients' experience and sophistication is apparent in their requests, which have gone from "Give me my data at the end of the project" to "Give me my data in machine-readable form" to "Give me my data in a form readable by my specific CADD and facility management systems." It is likely that sophisticated clients increasingly will demand this in the future. Many major organizations already utilize this information.

Databases from completed projects also form an extremely valuable historical record that is often called a *project history system*. As an organization repeatedly undertakes similar projects, it can often reuse substantial amounts of data from previous projects. Even when specific data cannot be reused directly, analysis of a number of previous projects yields valuable general information. Architectural firms are beginning to discover the benefits of establishing computer-based project history systems, which allow the setting down of intellectual capital into permanent, accessible form. Facility man-

agers will quickly make the same discovery as they begin to use automation systems.

Extremely large databases can be stored compactly, cheaply, and permanently on magnetic tape. Storing data on tape (from which reports, drawings, and summary decision graphics can be generated, as required) greatly reduces the amount of floor space that must be allocated to historical files and lowers the cost of maintaining such files.

SYSTEM INTEGRATION

It is technically feasible to automate a great many space-planning tasks. Economic success however, depends not only on having effective programs for automation, but on finding effective ways to integrate these programs into a true system.

Ad hoc introduction of uncoordinated programs is unlikely to be cost-effective: data preparation and update costs alone usually obviate such an outcome. Moreover, the relatively uneconomic utilization of hardware and labor and the disruptions of established ways of doing things are likely to wipe out all benefits. Instead the objective must be to establish a carefully designed, integrated system that contains interrelated and compatible database management software, application programs, procedures (including questionnaires and other standard forms), clearly defined standards, and uniform training.

The concept of a *unified database*—a single information source throughout the project—is highly significant. Simply transforming information from humanly readable forms (plans and specifications) to machine-readable ones (magnetic tape and disk files) is seldom economical unless careful thought is given in advance to three goals: (1) low redundancy in the database; (2) acquisition of or interface with information from other machine-readable databases; and (3) reuse of information. Low redundancy means that information should not be repeated in several places; if this goal is achieved, updates and changes (so frequent in the design and facility management processes) become easy. Reuse of information is critical to successful applications because data entry and update costs are so high, and because information often is needed at several points in the process.

For example, the information that results from analysis of group space needs (projected square footages and communication/adjacency requirements) also forms the input for location decisions (site plans, stacking plans, and block planning). Many database management systems provide tools to facilitate achieving these goals, but extreme care must be taken to design a data structure that is appropriate to the problem. A clear specification of need (including desired reports), followed by empirical "road testing" of a proposed system, is a good way to begin the investigation.

All but the very largest organizations must implement such a system piecemeal. The capital required for the necessary research and development work, staff retraining, and computer hardware simply is not available; and the prospective disruption of established office procedures is too great. Even where the developmental resources exist, the burdens of long-term system maintenance and enhancement are immense. Thus, a long-range plan for carefully phased implementation involving planned incremental acquisitions must be followed and is (in any case) most cost-effective.

OVERALL MASTER PLANNING

An interesting and important use of facility management systems is in developing long-range master plans. Usually the master plan identifies requirements for many years in the future; the time frame can be as short-term as one or two years, although it is almost always five or ten years and sometimes extends twenty or twenty-five years into the future. The long-range planning process should be completed to an appropriate level before annual or periodic plan-

ning and specific project implementation proceed.

The long-range planning process includes various preplanning activities, including analysis of alternatives and the preparation of outline reports documenting goals and objectives to be achieved, alternatives and options to be identified for analysis and future development, and the sequential courses of action leading to implementation.

The long-range planning process, potentially the most important aspect of space management, establishes an attitude of anticipation and action (as opposed to reaction) toward stimuli external to the facility management organization. While responding to day-to-day situations, outside or macroeconomic changes, and maintenance and rearrangement activities is clearly essential, the space management organization must also anticipate action (thinking down the paths that may have to be explored if certain events occur). Having a clear understanding of the long-range plan and using that plan as a road map for future activities are most beneficial.

Change is the only certainty over a long time period; and because the long-range master plan must be responsive to change, the plan should incorporate a high degree of flexibility. The master plan also should identify the limits defining the most likely range of future requirements. This suggests establishing a *planning horizon:* a set of minimum and maximum limits for space requirements, staff sizes, future operating budgets, and so on. A long-range plan is thus a road map with a series of branches and alternatives leading to a general point on the horizon.

Finally, the long-range planning process must allow updating by in-house personnel. Without this capability, the information in the system quickly becomes useless. Here, again, procedures are needed to identify the appropriate information capture points, to ensure that the data gathered is entered accurately. A database of inaccurate information is worse in many ways than no database at all: managers operating without a database at least know what they do not know.

Preplanning Activities

The most important part of the long-range planning process is sometimes termed *preplanning activities*. These activities are steps of the planning process that must be completed prior to initiating any space programming, space planning, or implementation project.

Preplanning activities include developing record data (such as as-built drawings, core drawings for new space, area assignment diagrams for existing conditions, personnel and equipment inventories, and preliminary projections of future requirements) and performing technical analyses of these data in order to evaluate available alternatives (for example, lease, relocate, remodel, construct, consolidate, or decentralize).

Most problems encountered in the course of the space management process and after a facility is occupied result from a lack of adequate attention to preplanning activities and from improperly prepared space programming data. In the absence of proper planning, there is often insufficient time available to conduct the required analyses and to explore every alternative fully. Space management personnel sometimes do not have the technical background necessary to complete those analyses. Further, even if adequate time and appropriate technical talent are available for the analyses, incomplete record data and space information may prolong the time necessary for analysis.

At the outset of many projects, accurate area assignment diagrams, record drawings, and core drawings are not available, and the space-user departments have relatively poor data on their current equipment and furniture inventories. Organization charts commonly are out of date, inaccurate, and unenlightening as to the true internal working relationships of the space user. Equipment and furniture inventories are gen-

erally compiled for entire organizational groups, rather than by specific assignment. In general, current record data serve administrative and budgetary purposes but are ill-suited to accommodate the informational needs of space planners.

Data on future requirements generally reflect hasty preparation and little analysis. Most often the information provided is far too general to be useful or is simply an assemblage of available raw data that may well be incomplete or unintelligible.

The overall success of a completed space action cannot exceed the quality of the weakest link in the chain of activities necessary to complete that space action. Certain preplanning activities, along with the preparation and analysis of user requirements, provide the foundation for all further study elements. If that foundation is not accurately analyzed and fully developed, the resulting solution may suffer.

Specific preplanning activities include the following elements:

- Space management and classification
- Analysis of alternatives
- Report preparation
- Strategy analysis

Space Management and Classification: The procedures utilized for area measurement and classification are fundamental to the development and maintenance of a comprehensive space management system. Two primary objectives are related to the measurement of occupied space: an accurate accounting of the facility resources occupied and maintained by all departments; and a means of evaluating space utilization efficiency.

In addressing the first objective, emphasis must be placed on measurement procedures that are absolutely uniform, in order to ensure that a compatible format of space accounting is achieved for all facilities, regardless of their size, location, or configuration. The second objective is useful to apply to any facility being considered for acquisition or to any space currently being occupied. To serve this need, accurate measurement of space (in terms of absolute usable area) is essential. The calculation of utilization efficiency should be based on occupied area per person, if procedures for life-cycle cost evaluation are to be applied.

Crucial to a space management system is the manner in which facility resources are classified by type and function and evaluated by use. The classification system should identify office, laboratory, residential, warehouse, and various special area types. Definitions adaptable to diverse interpretations and applications should be developed. Furthermore, this system should distinguish among the many types of office space. For example, differences between small, fixed (load-bearing), single-story, inflexible space and large, open, loft space should be noted. Assuming that all assignable office space has the same utility may lead to inappropriate conclusions in the analysis of specific space actions.

Secondary to the description component of the classification system is a space quality indicator. The availability of this type of data to the facility planner (at the level of master-plan study) and to the space planner (at the level of specific facility renovation) is important for feasibility studies and estimation of preliminary construction costs.

Analysis of Alternatives: At this point in the process, all record data have been provided and user space requirements will have been projected. It is then necessary to analyze alternative potential solutions, such as the following:

· Leasing new space, and relocating for a short or long-term
· Remodeling and possibly expanding in currently leased space
· Splitting operations into two or more locations
· Adopting centralized (versus decentralized) services
· Relocating to owned facilities
· Identifying temporary ("holding") situations, until more appropriate long-term solutions are feasible
· Developing a new facility specifically to satisfy space user requirements

· Developing a building that will provide space for the user

In order to perform a comparative analysis of these alternatives, the planner must first identify all quantifiable costs. Such costs might fall under the following heads:

· Remodeling
· Land acquisition
· New facility construction
· Relocation
· Leasing
· Furniture procurement or refurbishment
· Facility operation and maintenance costs

In addition to these quantifiable variables, certain potentially unquantifiable variables must also be considered. Variables in the latter group include the following:

· Demographics and location
· Relationships with other departments
· Impacts of changes on department or space user efficiency
· Security
· Safety
· Provision for future growth
· Need for flexibility

In the final analysis, one of three general solutions will be indicated: remodeling/relocation within owned space; leasing new space; or building new space. If a remodeling activity or relocation to an owned building is indicated, the planning process can proceed with technical space-programming activities. If a leasing action is indicated, development of a space-soliciting document will be necessary. If new construction is indicated, the subsequent planning process will commence with prearchitectural activities.

Space actions should be completed in a cost-effective manner that emphasizes the minimization of initial costs and (just as important) utilizes techniques for analyzing present-value life-cycle costs. Analysis procedures and budgeting and funding systems must be developed to accommodate this process. All variables—

both quantitative and qualitative—should be considered in the analysis of alternatives.

Report Preparation: Each year, an annual plan should be prepared and five-year and long-range (fifteen- to twenty-year) plans should be updated to the degree required. This annual plan should be prepared before departmental budgets are submitted and certainly before those budgets are finalized for the coming year. The one-year and (possibly) five-year capital improvement programs should highlight specific line items that must be included in the annual budget. Line items reflecting necessary remodeling and rearrangement activities or concentrating on funding for long-range planning, design, and fees associated with preliminary architectural activities should be highlighted.

The long-range plan requires that program statements be developed reflecting service requirements by all departments. The service programs involved must be translated into space programs, which in turn (after comparison to existing space resources) must be converted into specific leasing, construction, and remodeling statements.

Strategy Analysis: A most important preplanning activity—one that should precede any implementation task—is the development of a long-range strategy. As part of the master-plan process, strategy analysis may focus on straightforward issues such as which space to lease and on more complex issues such as centralization versus decentralization, relocation of offices in a new community, or analysis of lease versus ownership as the more appropriate space acquisition strategy.

Analysis of alternative space acquisition or implementation strategies can be reasonably simple (if the organization's policies and procedures have already dealt with the issues at hand) to complicated (if a very stable historical environment has not changed in scale or location and does not address the issues at hand).

Various strategies that could be considered in the formative stages of the project might include the following:

- Leasing or acquiring an existing facility, and renovating it
- Acquiring a site, and building on it
- Centralizing or decentralizing operations
- Performing alternative locational studies
- Using private offices versus developing open office planning concepts
- Exploring alternative interior development and planning concepts
- Undertaking phased or one-time construction
- Subleasing expansion space, or warehousing vacant space
- Weighing the relative merits of suburban and urban locations
- Designing, constructing, and implementing versus securing a development consultant versus engaging an architect and then a contractor
- Making use of fast-track or competitive bid construction

Decisions must be made with respect to these important acquisition and implementation strategies, and answers must be supplied in response to these strategy issues before implementation activities can proceed. Consultants who are familiar with the issues and have previously developed master plans for space management systems and facility development strategies can assist the space user in exploring the alternatives and determining appropriate strategies to adopt.

Long-range Planning Tools

Most tools used in performing long-range master planning are similar to those used in performing programming for group space needs. The primary difference is in the level of detail. In space requirements programming, the gathering of detailed information (such as number of desks required, individual employee names, and employee locations) often proceeds from the bottom up. In long-range master planning, broader information (such as generic equipment counts, number of job titles, and square footage of sites) is projected from the top down.

Many computer-based tools have been used for these applications, including scores of architectural programming systems such as Morganelli-Heuman's DATMAN (Data Management), Skidmore, Owings & Merrill's SARAPI (Storage and Retrieval of Architectural Program Information), and the Computer-Aided Design Group's UNI (User Needs Information) and Facility Requirements Programmer (fig. 6-3). Most of these systems can trace their genealogy to either DATMAN or SARAPI.

Periodic Update/Planning

Within the context of the overall master plan, many space management organizations develop periodic (annual or semiannual) updates and refinements. Since the long-range planning process focuses on a road map and a general direction of future development (to a point where space requirements exceed space resources and adjustments are necessary), a more focused master plan that identifies immediate needs is extremely helpful. Periodic master plans may focus on a particular project or on a series of projects that require implementation.

The periodic planning process is, in a sense, a stepping stone in the overall planning process from the conceptual level presented in the long-range plan to the concretely detailed level addressed in specific project implementation. Periodic planning serves two major purposes: it updates and monitors record data, and it identifies specific projects that are to be implemented.

As the process moves closer to project implementation, the following tasks become necessary:

- Develop budgets for projects.
- Secure authorizations and funding appropriations.
- Develop project schedules, and monitor and review needs to ensure completion of the project on time and on budget.

Specific Project Implementation

The planner's decision to implement one specific project over another should be based on understanding the current problems and solving (or not solving) each.

Preliminary Space Programming: Before attempting to prioritize specific project requests, the planner must update and verify data that identify what space actions are needed. By completing this intermediate step, the planner gains a better understanding of each request, which permits a more solid basis for ranking the needs. In addition, this preliminary space programming phase acknowledges changes in data and/or needs.

The first step of preliminary space programming should include itemizing current staff levels by name, title, and organizational level within the group. This allows calculation of existing area factors and correlation of positions and workspace, thereby facilitating the evaluation of existing conditions in comparison to the workstation standards set forth in the long-range planning process.

Projected future staff levels should also be adjusted to reflect changes that may have occurred in the period that has elapsed since prior planning. The revised staff projections, coupled with appropriate workstation standards and special area requirements, identify shortfalls of space for each department that has requested action. Other criteria may be established, as needed, to rank requests for space action. For example, the need to provide additional space to one particular department may be perceived as more important than the need to do so with respect to another department—notwithstanding the conclusions drawn from the primary consideration. Procedures, as they are established, should include room for these sorts of additional considerations.

The preliminary space programming process offers a simplified method of evaluating and ranking requests for space action and of up-

Projection Analysis Report 15: Area Analysis; Personnel by Standard

This report is an analysis of Report 14: Summary of Personnel by Position Standard which presented Area requirements for individual Position Standards. Its layout is identical to that of the Summary Report but the Position Standard Area requirements are shown as percentages of the Unit or Activity's total Area requirements. This type of analysis report allows you to spot the Positions that are large contributors to the total demand for floor Area within individual Organization Units or Activities.

SUMMARY NEEDS   Organization: 7200   Marketing Division

POSITION STANDARD		QUANTITY				AREA AS % OF TOTAL AREA			
ID	NAME	06/85	06/86	06/87	06/90	06/01/85	06/01/86	06/01/87	06/01/90
PERSONNEL									
ADM-A2	Admin standard A2	3	3	3	3	5.35	5.31	5.19	4.85
CLR-C1	Clerical standard C1	3	3	3	3	2.34	2.32	2.27	2.12
MGR-B1	Manager standard B1	6	6	6	7	10.70	10.61	10.39	11.31
MKTG-A1	Marketing standard A1	10	10	11	12	11.14	11.05	11.90	12.12
MKTG-A2	Marketing standard A2	8	8	8	9	8.91	8.84	8.66	9.09
	Projection	27	27	28	31	38.44	38.14	38.42	39.48
	Circulation	12%	12%	12%	12%	4.61	4.58	4.61	4.74
	Subtotal				...%	43.05	42.72	43.03	44.21
EQUIPMENT									
	Projection					22.28	22.79	22.98	23.36
	Circulation	5%	5%	5%	5%	1.11	1.14	1.15	1.17
	Subtotal				...%	23.40	23.93	24.13	24.54
SUPPORT SPACE									
	Projection					22.23	22.05	21.59	20.14
	Circulation	10%	10%	10%	10%	2.23	2.21	2.16	2.02
	Subtotal				...%	24.46	24.26	23.75	22.16
MARKETING DIVISION SUMMARY									
	Projection					90.91	90.91	90.91	90.91
	Design factor not used	0%	0%	0%	0%	0.00	0.00	0.00	0.00
	Subtotal				...%	90.91	90.91	90.91	90.91
	Allowance for uncertainty	10%	10%	10%	10%	9.09	9.09	9.09	9.09
	Total				...%	100.00	100.00	100.00	100.00

Figure 6-3. Facility requirements programmer. (Courtesy of Computer-Aided Design Group)

dating the information to be used in further analysis. Detailed analysis should begin after specific projects have been earmarked for implementation.

Case Manager and the Case Team: A case manager is selected to handle each request for space action. The case manager's responsibility is to head a project team and ensure that all aspects of the assigned project are completed satisfactorily, within the established budget and time schedule. An individual may serve as the case manager for one large project or for several small projects simultaneously.

For each project, regardless of its size, a case team is created. The case team is composed of all of the experts who are expected to perform significant work on the project at one time or another. Typically, the team includes (at a minimum) the following members:

- Case manager
- Acquisitions specialist
- Space planner, with expertise in programming, layout, and design (or separate individuals with such expertise)
- Construction management specialist
- Buildings management specialist
- Client representative

The team is organized at the outset of the project and remains operational until the project is fully completed. The team periodically reviews the status of the project and gives each member the opportunity to make suggestions or offer opinions. Although it is convenient for team members to be located in the same area, they can communicate entirely by memorandum and telephone, if need be. Team members are likely to serve on more than one team at a time.

Prioritization of Requests for Space Action: Organizations should continually monitor their space needs. When a space problem arises, the department involved should submit a *preliminary request* to the space management unit, identifying general requirements on the basis of calculations from established planning factors

(rather than from detailed analysis). The space management unit then uses objective *ranking criteria* to rank these preliminary requests. The criteria include functional (department-benefit and organizational-benefit), public-benefit, and cost factors.

Functional factors measure the existing necessity or resulting benefit of a requested change to the individual department and the overall organization, in terms of what needs are satisfied (adjacency, services, equipment, and so on). Similarly, the necessity or benefit of the change to the general public is evaluated, where appropriate. Finally, costs and/or savings attributable to implementation are estimated. All criteria should be evaluated in dollar or numerical terms whenever possible, and they should take into account life-cycle costs and savings. This approach permits a serial ranking of the requests and gives preference to projects that generate a cost savings or that minimize overall costs. When all requests have been ranked in order by their cost/benefit ratio, the space management unit informs each user department of its ranking and of the estimated future time when its requests will be processed.

User departments with high-ranking preliminary requests are asked to enter the second cycle of the request phase: preparation of detailed requirements. Knowing that their request will soon be processed, these user departments can more readily be convinced to make the significant effort necessary to identify their space requirements in detail. Submission of the user's detailed space requirements completes the two-step request submission process.

Space Management's Managerial Role: If the space management unit is to deal successfully with such problem areas as time schedules, a backlog of requests, the quality of its services, and allocation of limited resources, it must emphasize *managing* the process rather than doing the process. It must think as a manager (coordinate and direct), not as a technician (perform some concrete part of the project). The process must be viewed as a process, rather than as a series of connected but independent work tasks.

As has been discussed previously, one of three general solutions will be identified for future space requirements: new construction, leasing, or remodeling existing space. Should new construction be indicated to accommodate future space requirements, the next step in the planning process is to complete prearchitectural studies. Should a leasing action be indicted as the best way to accommodate a user department, the next step in the planning process is to prepare a space solicitation document. Finally, should remodeling activity or relocation to an owned building be indicated as the most appropriate means of satisfying space requirements, the planning process proceeds according to the technical space programming activities that the project raises.

The preparation of the space solicitation document (when a leasing solution is adopted) sets the tone and establishes the context for the balance of the space planning and occupancy program in this case. If the document is not properly and comprehensively prepared, the space user may find upon occupancy that many specific functional requirements are not satisfied, and future growth and rearrangement may be restricted. The space solicitation should include the following features:

- A standardized format
- Functional performance criteria
- A framework for identifying specific user requirements
- An evaluation technique to allow selection of the most appropriate lease space for utilization

Once the annual plan is complete, and the space requests appropriate for accommodation in the current fiscal year have been selected, the active stage of specific project implementation begins.

An interesting departure from these approaches to master planning is the Computer-Aided Design Group's Facility Master Planner.

This is the first commercial attempt to provide the facility executive with a computer-assisted program for gaming and improving multiple scenarios simultaneously. This is closely akin to the real-world needs of facility master planners. The program uses the concept of snapshots that take into account such variables as cost, move time, facility resources, and facility needs (for buildings, space, functions, and so on). The snapshots include information that can be extracted from other facility management system databases maintained by the firm, such as the resources and needs inventories. Examples are shown in figures 6-4 and 6-5.

SCOPE OF SERVICES OUTLINED

Jim Steinmann has noted that many of the activities required for the completion of a space-planning and interior design study are also included within the scope of a master planning study. The following outline, developed by the firm of Steinmann, Grayson, Smylie, identifies activities normally associated with one or the other of these projects. The outline, a slightly adapted form of which is given in this section, is introduced by Steinmann as follows:

> The activities are divided into five phases, which detail tasks from the start of a project (orientation) to the conclusion (implementation). Master planning requires the development of a comprehensive database and long-term strategy analysis, as in Phases A and B. Phase C, "Concept Development," normally is reviewed and updated on a more frequent basis, referred to as *periodic planning*. The final two phases outline specific project activities and include finalization of concepts and their implementation. Design studies are associated with specific projects (as detailed in the final three phases), although their completion requires the availability of information developed in the first phases.

Phase A: Requirements Analysis

1. **Orientation:** Review current organization of departments, divisions, and existing space.

Summary of Snapshot Group Allocations (SNAP2)

This report shows Total Need, Total Allocation, and Remaining Need for each Group-Space Category. The totals are shown Snapshot by Snapshot and represent Group-Space Category needs, allocations and remaining needs at all Location-Space Categories combined.

Facility Master Planner
Snapshot Report 2 (SNAP2): Summary of Snapshot Group Allocations

Scenario	Snap	Group	Group Name	Sp Cat	Needed Area	Alloc Area	Remain Need
FYE1990A	1	10000	Production	MFG	400000	400000	0
FYE1990A	1	10000	Production	OFF	40000	40000	0
FYE1990A	1	11000	Sales	OFF	50000	50000	0
FYE1990A	1	8000	Development	OFF	70000	60000	10000
FYE1990A	1	9000	Research	LAB	80000	80000	0
FYE1990A	1	9000	Research	OFF	30000	30000	0
Snapshot Total					670000	660000	10000

Scenario	Snap	Group	Group Name	Sp Cat	Needed Area	Alloc Area	Remain Need
FYE1990A	2	10000	Production	MFG	590000	400000	0
FYE1990A	2	10000	Production	OFF	50000	50000	0
FYE1990A	2	11000	Sales	OFF	60000	60000	0
FYE1990A	2	8000	Development	OFF	80000	60000	20000
FYE1990A	2	9000	Research	LAB	90000	80000	10000
FYE1990A	2	9000	Research	OFF	40000	30000	10000
Snapshot Total					910000	680000	40000

Scenario	Snap	Group	Group Name	Sp Cat	Needed Area	Alloc Area	Remain Need
FYE1990A	3	10000	Production	MFG	1060000	1060000	− 190000
FYE1990A	3	10000	Production	OFF	60000	60000	0
FYE1990A	3	11000	Sales	OFF	70000	70000	0
FYE1990A	3	8000	Development	OFF	70000	70000	0
FYE1990A	3	9000	Research	LAB	110000	110000	0
FYE1990A	3	9000	Research	OFF	50000	50000	0
Snapshot Total					1420000	1420000	− 190000
Scenario Total					300000	276000	− 140000
TOTAL					3000000	2760000	− 140000

Figure 6-4. Facility master planner. (Courtesy of Computer-Aided Design Group)

Summary of Snapshot Locations Report (SNAP5)

This report shows Total Available Area, Total Allocated Area, and Total Remaining Area at each Location-Space Category. Preassigned Location Goals are also shown. The totals are shown Snapshot by Snapshot.

Facility Master Planner
Snapshot Report 5 (SNAP5): Summary of Snapshot Locations

Scenario	Snap	Loc	Location Name	Sp Cat	Location Available Area	Location Alloc Area	Location Remain Area	Goal
FYE1990A	1	HQR	Headquarter Building	OFF	110000	100000	10000	....
FYE1990A	1	LAB1	Laboratory Block One	LAB	70000	60000	10000	....
FYE1990A	1	LAB1	Laboratory Block One	OFF	50000	30000	20000	....
FYE1990A	1	PLT1	Plant One Building	MFG	400000	420000	−20000	....
FYE1990A	1	PLT1	Plant One Building	OFF	50000	50000	0	....
Snapshot Total					680000	660000	20000	

Scenario	Snap	Loc	Location Name	Sp Cat	Location Available Area	Location Alloc Area	Location Remain Area	Goal
FYE1990A	2	HQR	Headquarter Building	OFF	70000	70000	0	....
FYE1990A	2	LAB1	Laboratory Block One	LAB	70000	70000	0	....
FYE1990A	2	LAB1	Laboratory Block One	OFF	130000	70000	60000	ON
FYE1990A	2	PLT1	Plant One Building	MFG	400000	410000	−10000	....
FYE1990A	2	PLT1	Plant One Building	OFF	50000	60000	−10000	....
Snapshot Total					720000	680000	40000	

Scenario	Snap	Loc	Location Name	Sp Cat	Location Available Area	Location Alloc Area	Location Remain Area	Goal
FYE1990A	3	HQR	Headquarter Building	OFF	70000	70000	0	....
FYE1990A	3	LAB1	Laboratory Block One	LAB	110000	110000	0	....
FYE1990A	3	LAB1	Laboratory Block One	OFF	210000	120000	90000	ON
FYE1990A	3	PLT1	Plant One Building	MFG	1720000	1060000	660000	....
FYE1990A	3	PLT1	Plant One Building	OFF	80000	60000	20000	....
Snapshot Total					2190000	1420000	770000	
Scenario Total					3590000	2760000	830000	
TOTAL					3590000	2760000	830000	

Figure 6-5. Facility master planner. (Courtesy of Computer-Aided Design Group)

2. **Prior documentation:** Review previous studies, as-built drawings, and existing data.
3. **Inventory of space and furniture:** List current space allocation and utilization, identify special areas, and note condition and adequacy of space and facilities.
4. **Formulation of future business plans:** Analyze current strategic and functional business plans and workload indicators at corporate level, and project future business volumes.
5. **Database analysis:** Develop questionnaire for collecting historical data and data on future staffing, adjacencies, workload indicators, special needs, overall current and future space requirements, and current improvement potential.
6. **Adjacency requirements:** Develop bubble diagrams and detailed analysis of adjacencies.
7. **Special area studies:** As above (item 3), study requirements for special equipment areas (central duplicating area, testing room, data processing center, and so on).
8. **Block layouts and expansion planning:** Diagram area assignment blockouts and stacking plans, identifying each department and division and providing for sequential expansion accommodating future growth. Identify areas for private offices, conference space, and support activities.

Phase B: Strategy Analysis

1. **Alternative locational considerations:** Examine potential costs and benefits of alternative locations in light of adjacencies previously identified (Phase A, item 6), including detailed account of actions needed to make such moves.
2. **Development of master-plan strategy:** Identify alternative scenarios related to organizational growth over (at least) ten years. Alternatives should consider circumstances such as centralization versus decentralization.
3. **Financing alternatives:** Identify alternative financing techniques, their limitations, their benefits, and their potential impacts on scheduling.
4. **Interim report and review:** Review final comprehensive database and all work completed to date, to permit user interaction and assessment (approval or recommendation of change).

Phase C: Concept Development

1. **Workstation requirements:** Develop workstation standards, adjusting where necessary to accommodate unique requirements. Develop sample workstation schematics.
2. **Analysis of systems furniture feasibility:** Develop life-cycle cost analysis for conversion of appropriate space to systems furniture. Evaluate systems in order to select most functional and economical application.
3. **Budget:** Develop preliminary project implementation budget.
4. **Schedule:** Develop preliminary implementation schedule, showing decision and action points but avoiding excessive detail.
5. **Interim report and review:** Facilitate user input and review.
6. **Finalization of workstations:** Finalize all functional workstations, based on selected interior planning concept and preliminary budget.
7. **Preliminary space plans:** Develop preliminary space plans, locating all workstations, equipment, and offices by group or department.
8. **Interior design program:** Design and secure approval of plans for design concept, color palettes, floor and wall coverings, paint colors, and special design features. Develop detailed furniture plans, translating workstation standards into scale drawings.

Phase D: Concept Finalization

1. **Working drawings:** Prepare working drawings, including demolition plans, construction plans, reflected ceiling plans, telephone and electrical plans, special details, and furniture installation drawings.
2. **Furniture requirements listing:** Itemize and describe all existing and recommended new equipment and furniture.

Phase E: Concept Implementation

1. **Implementation schedule:** Develop detailed schedule and coordination of outside subcontractors, utilities, and others.
2. **Final budget:** Itemize estimated costs for dem-

olition, construction, finishes, furniture, furnishings, and accessories.

As has already been noted, Steinmann's outline deals with the activities involved in space planning and interior design projects, and many of these activities also are required in overall master planning.

There are parallels (most notably time relationships) between the planning activities identified earlier (space management and classification, analysis of alternatives, report preparation, and strategy analysis) and those outlined by Steinmann (see fig. 6-6). Phases A and B of Steinmann's outline generally occur early in a project, while phases C, D, and E are specific to an implementation. Note that Steinmann here is describing activities that are project specific, whereas earlier we were discussing overall planning. Thus, in Steinmann's outline the term "strategy analysis" refers to analyzing the possible strategies for solving a specific space planning/interior design problem; "strategy analysis" as used earlier refers to possible strategies for overall master planning solutions. Both are

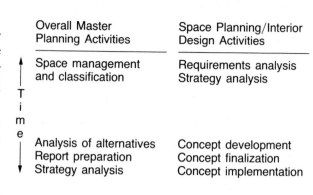

Figure 6-6. Time relationships of overall planning and project-specific activities.

planning problems and some of the methodologies (such as information gathering and analysis, production of reports showing concepts and alternatives, and analysis of the alternatives) are similar except for differences in scale and elements of the problem.

SEVEN

MANAGEMENT OPERATIONS

The final subsystem of a comprehensive facility management system is management operations. Building management operations comprise maintainance, decision-support, and financial management activities. Figure 7-1 is a simple way to depict some of the forces affecting an organization's facility management operations.

PLANNING MANAGEMENT

In order for an organization to manage future facilities requirements, two basic actions are necessary: the (floor) area required at the future date must be determined; and the question of whether the future area requirements can be accomodated in available built space must be resolved.

Future Floor Area Requirements

Establishing future floor areas involves developing an inventory over time of items that consume floor space and indicating how these items are likely to change quantitatively and qualitatively over time. The amount of future floor area established depends entirely on the meth-

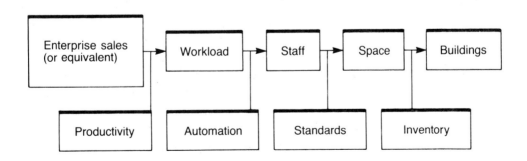

Figure 7-1. Forces affecting facility management operations.

ods used to project or validate the need for present floor area. Consequently these methods must be believable, simple to derive, and easy to manipulate. They should be presented in a logical manner, with no ambiguity as to how the present area is to be transformed into the future area.

In practice, these results are often expressed as percentages, since they represent compound annual rates of change that are often determined from growth projections for the overall organization. The sources should be as varied as possible: supplied by the user, derived from past facilities trends, obtained through specialized model (economic, production) analysis, and so on.

Future floor areas can then easily be established or projected to a future date, using the appropriate method(s). Ideally, the process of projecting should allow a variety of approaches: a top–down approach by means of which groups of items consuming floor area (such as an entire division or a department) can be anticipated or manipulated as a whole; a bottom–up approach by means of which the growth of individual items of floor area consumption—such as standard workstations, equipment, and filing—can be monitored; and representative combinations of the two. A comprehensive projection system thus allows a mix of both approaches. Either "pure" approach may serve as the chief means by which future target figures are generated, with the alternate approach(es) used for exceptions. As detailed results are generated, they can be compared with the established or projected top–down results to determine if the intent of the master plan is being realized.

Allocating Projected Area to Built Space

Projecting future floor area requirements establishes one aspect of facility management— the floor area that will be required—but it does not indicate how the future area requirements are to be accommodated in present or future facilities. This is the point of compromise between needs and resources. Comparisons among existing areas and future alternatives can be made to highlight differences, but the fundamental facilities question to be answered is "can the facilities handle the future requirements?"

In order for the planner to answer this question, future area requirements have to be placed in the anticipated available space. Doing this in accordance with a bottom-line or top–down approach tells the planner whether the original assumptions are even in the ballpark. Detailed department-by-department analysis also is required, however; it is not sufficient merely to substitute a department's future area requirement for its present area requirement. Additional important questions must be answered, too.

For example, Department A has expanded its floor area requirements by 50 percent. Approximately 25 percent of this can be accommodated on the floor it currently shares with Department B. But B also is expanding, and it could also use the available space. Should the facility manager move A and use the floor space for another department (if so, which department?) or move B and find a use for the extra space? Whichever one is selected, where should it be moved? What about the new space on the nineteenth floor that is being offered by the landlord? What are the lease, cost, time, and construction implications? Is it really desirable to move A, given that it has two computers that were very expensive to install?

The problems just raised can be solved by using allocation algorithms, as discussed earlier. There is, however, a need for a higher-level, master-planning function. This is the three-dimensional chess game of titrating the three primary variables in any project: cost, time, and quality. A truism of project management states that you can never attain optimal levels of all three simultaneously: as you push one, it adversely affects one or both of the others (fig. C-35 of color insert).

CHANGE MANAGEMENT

Occupied space continually changes in response to an organization's requirements. The reasons for changing a facility are to ensure (among other things) that needs are met adequately over time, that costs are controlled, and that waste and duplication are minimized. Achievement of these aims implies systematic management of the facility change process (Fig. 7-2).

Systematic management of facilities changes involves the following steps:

1. Identify and arrange facility change projects in order of importance, cost, and so on.
2. Plan and manage each project in terms of budget allocation, tracking, and reporting (contract administration); allocate personnel and resources; track and report progress.
3. Perform detailed space programming, interior design, and architectural design.
4. Maintain records of facilities in as-built form.

Identifying Facility Change Projects

Virtually all facility changes must be accomplished by certain dates. For organizations undertaking the changes on the strength of limited personnel resources (most organizations, in other words), projects should be scheduled in order of importance and in order of implementation. Overall project planning maximizes the chances of getting the required changes done on schedule and minimizes the resulting waste of personnel resources, money, and time.

Project Planning and Management

Once change projects have been identified, each must be planned and managed systematically. The planning, design, and construction phases of each project should have individual budget allocations, time schedules, and manpower allocations. Expenditures in terms of budgets, time, and personnel should be traced and re-

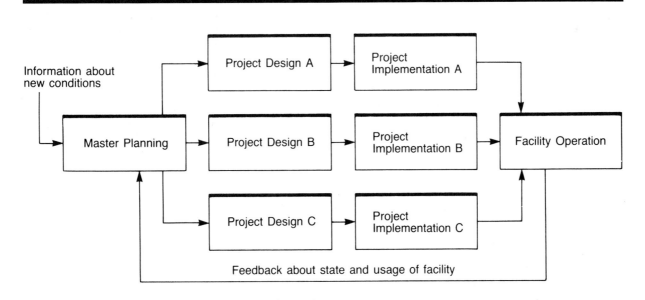

Figure 7-2. Cyclical process of facility management. (Courtesy of Computer-Aided Design Group)

ported on both a spent-to-date basis and a percentage-to-complete basis. Resource allocation should be performed and periodically adjusted.

Hundreds of project management systems are commercially available. Most are automated, utilizing CPM (Critical Path Method) and/or PERT (Program Evaluation Review Technique) algorithms. Such systems need to be reviewed in accordance with the criteria applied (elsewhere in this book) to the selection of other automated facility management systems. These criteria include capacity for database integration with other facility management applications, ease of training and use, vendor standing and support, hardware/software operating environment, and suitability for the technical task to be performed.

Detailed Space Programming

Changing a facility invariably requires developing a detailed space program database—that is, carrying out a reprogramming exercise. This activity goes a long way toward ensuring that the completed design will provide the required facilities adjustments in a manner highly responsive (and satisfactory) to the users of the space. Since it functions essentially to provide updates to the existing space program, it need not be an inordinately time-consuming activity.

Maintaining As-Built Facility Records

An organization's knowledge of the present state of its facilities depends largely on whether it maintains facilities information in as-built form or not. Noncurrent facilities information severely limits (and can even prevent) effective management—present or future. Again, procedures (for recording each change on completion) are the guarantor of information currency.

SPACE RESOURCES MANAGEMENT

While it is important to be efficient and economical in managing space occupied by the space user, managing the inventory of space owned by the space user but leased to others is also a significant endeavor. Managing this space effectively can maximize its utility, provide expansion capability, and ensure that vacancy factors are kept as low as possible. The subsections that follow discuss the salient characteristics of a portfolio management system.

Owned Space

Managing owned space resources requires careful tabulation and allocation of costs related to different occupancies and organizational units. More important, an effective space management organization must carefully monitor the level of space utilization (the net area allocated per person). The goal here is to minimize the cost of operating and maintaining space, and to project and anticipate future space needs for expansion or contraction accurately.

Managing owned space is somewhat less complicated than managing leased space because the available space resources are relatively more stable, the level of inventory is less likely to change quickly, and the costs sometimes are more controllable than in a lease situation. Nonetheless, the continuing need to anticipate future requirements for the various occupancies and the need to develop a comprehensive plan of action to satisfy those requirements are of paramount importance. Projected needs are more difficult to gather (and verify) from occupants than existing needs are. The managers of these space resources must rely on tools and procedures (as discussed elsewhere in this book), including development of a long-range facility master plan and (often) development of space programs and prearchitectural specifications.

Leasing and Subleasing

Large organizations that own buildings or occupy large amounts of space frequently find it desirable to lease space to others in order to generate income or to protect expansion capabilities. Managing lease space involves significant day-to-day activities on the part of the property manager. In many cases, these duties are subcontracted to others, such as professional building management and/or leasing agents—especially when the owner's primary business is something other than property management. The owner in this instance is responsible for ensuring that the expansion and rearrangement strategies intended to support internal operations are suitable in the context of the opportunities to take over lease space.

This section is not a discussion on how to be a landlord. That subject is covered extensively in the literature (publications of the Building Owners and Managers Association are a good source of information). Here the discussion is limited to facility management systems applications.

The property manager must have a clear, accurate, and up-to-date inventory of all space that is leased to others (out-leases) and of space that is leased from others in buildings not owned by the user (in-leases). The inventory should accurately identify the following items:

• The location of the lease
• The terms and conditions of payment
• The renewal options
• The capital construction value
• The annual depreciation
• The occupant/occupancy load, space utilization (percentages of open and closed office, warehouse, laboratory, light and heavy manufacturing, and so on), and efficiency
• The attributes (such as parking provisions)
• The business terms (such as rights of first refusal, expansion clauses, buyout provisions, rental escalation formulas, and so on)
• The net and rentable areas
• The expenses incurred in conjunction with that particular space

These and other data should be maintained through utilization of a computer-based system for larger-scale applications.

A very wide variety of property management software exists, most of it competent. The most difficult features to find in property management software systems are the very ones needed for general facility management software: procedures supporting the software (usable with or without a computer); and integration of data from the property management system with data from other facility management system components. Many computer-based systems include formulas for determining rent increases and cost allocations for each particular space; such formulas also provide a simplified means of cost-accounting control—an important element in the cost-effective management of space.

The Computer-Aided Design Group has diagrammed the process of facility management as shown in figure 7-3. Jim Steinmann has done so as shown in figure 7-4.

Another highly desirable process to implement is an online automated system to project future space inventories that will become available as a result of lease expirations. Such systems identify all subleased space, reflect the expiration date of each sublease, and thereby anticipate the loss of space resources in any space occupied by the organization.

Forecasts of changes in lease structures, schedules for the expiration of leases, and other red-flag conditions can be automated. Monthly reports should be available to indicate when action needs to be taken on a decision to renew a sublease, to terminate it, or to begin negotiations for securing new space to replace space whose lease is terminating. In addition, such systems can help the user determine whether or not to conduct negotiations to extend occupancy in a leased facility. Often—particularly if new construction is contemplated—management may require three to five years' lead time to anticipate what to do and to develop a strategy for replacing or supplementing leased spaces.

In managing subleased space, decision points must be red-flagged some nine months to two years prior to expiration of a sublease, so proper planning can proceed on acquiring a new tenant, determining which department will occupy the

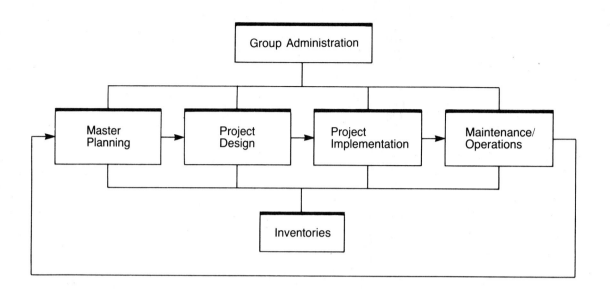

Figure 7-3. Facility management process. (Courtesy of Computer-Aided Design Group)

space, preparing appropriate implementation plans and specifications, and so on.

Many leases include rent-escalation formulas related to the cost of operating and maintaining space, to a consumer price index (CPI), or to other reference points. These vary considerably, and even a term such as *consumer price index* involves different locations, methods of cal-

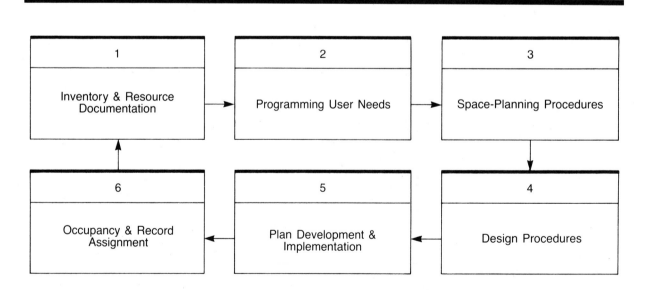

Figure 7-4. Space management process. (Courtesy of Steinmann, Grayson, Smylie)

culation, and time periods covered in different organizations.

Often, different formulas for identifying how annual rent increases or periodic changes are calculated can be automated and will accept information from the cost allocation subsystem of an integrated facility management system (or from other property management software or systems developed to manage a building). Automating in this way expedites the process and adds a level of accuracy to the development of modified rent structures.

FINANCIAL MANAGEMENT

Financial management is obviously of prime importance to any organization. It therefore becomes essential to design and implement accounting systems and to develop modern analytical techniques capable of accommodating the complex aspects of contemporary real-estate operations. Quantitative tools and techniques applicable to all aspects of real-estate investment (financing, marketing, valuation, risk/return analysis, property development, and property management) are available.

Perhaps the widest range of software systems available to property managers (appropriately) deals with financial management. Hundreds of computer programs are commercially available for operation on everything from personal computers to mainframes. Although description of these is beyond the scope of this book, many sources are listed in the appendix. The subsections that follow provide brief outlines of some factors involved in facility accounting.

Real-Estate Accounting

Property cost accounting has become increasingly important in recent years. One major cause is tax legislation and the rules of other government agencies; another is inflation, which makes measuring financial results more difficult and complicates the tasks of planning and controlling business operations. The periodic swings of real-estate markets, in concert with changing lease contract standards, mandate that organizations institute procedures for accurate accounting over time. The growing availability of computers and suitable software systems has facilitated the development of more useful financial reporting systems and techniques of analysis.

Financial Versus Managerial Accounting: The traditional distinction between financial and managerial accounting is especially significant in the field of real estate. Financial accounting involves preparing financial statements for individuals and entities outside of (or supervisory to) the organization; one example is external financial reporting. Managerial accounting, on the other hand, involves providing internal financial reports to aid in management, operations, and analysis (of such things as new investment proposals, decisions to keep or sell, and "what to do"). Managerial accounting may be thought of as operations accounting.

Specifics of Financial Accounting: To prepare financial statements, the person responsible must have access to an accounting system that is designed to record and accumulate financial data on a daily basis; in short, a bookkeeping system is required. Traditionally, such a system is built around a set (or chart) of accounts within which entries are made to the accounts affected by transactions during the period. Complex formulas are often necessary in allocating transaction costs to tenants or departments, and in many cases bills must be prepared. The balances in the system can be drawn off periodically for use in preparing financial statements. The nature of the statements depends on many factors, including the principal line(s) of activity (for example, rental property ownership, lot sales, or property development), the legal form of organization, and the legal form in which the enterprise is held.

Specifics of Managerial Accounting: Despite recent attempts to improve them, existing guidelines on the public reporting of financial data often result in unrealistic financial statements. As a consequence, alternative financial statements prepared for internal management's use take on greater significance. Moreover, the needs of internal management include much additional (optimization and operational) information. These reports of real-estate activities generally recognize the importance of cash-flow performance, as well as of traditional net income performance. They also attempt to take into account the impact of inflation (both real and market) and the action of other sources of change in property value, such as external and location-specific changes.

Unique forms of ratio analysis also are required to identify the profitability and risk features of real-estate investments: return on investment (ROI), gross income multipliers, return on equity (ROE), pretax and after-tax income, leverage, debt service coverage, and so on.

An important managerial accounting function consists of projecting future results for an investment in an income-producing property. The forecasting of future results is undertaken by the investor, as an element in assessing the potential returns of a proposed investment. Estimates also are made by the owner of an income-producing investment, in order to assess expected future income and expenses. The basic profit-planning tools include the pro forma statement, future profit and leverage ratios, and break-even occupancy analysis. These techniques are widely used in practice because they offer a straightforward approach to measuring the principal profitability elements for most real-estate projects.

Other Aspects of Real-Estate Accounting: Except in the case of raw land projects, large annual depreciation amounts are a feature of real estate as an investment vehicle. The manner in which depreciation charges are computed has a major impact on financial statements and helps determine the final amount of taxable income and the project cash flows. The laws regarding depreciation and deductibility of interest and expenses have changed periodically and significantly, and more changes may be expected.

Investigating future alternative investment scenarios requires the use of two basic approaches to depreciation (composite and component), as well as the principal methods of determining depreciation charges (straight line, sum of digits, and declining balance). Various assets—by type (buildings, FF&E [furniture, fixtures, and equipment], and so on), length of ownership, and date of acquisition or disposition, among others—are depreciated differently.

Increasingly sophisticated mathematical techniques have become a feature in the various aspects of the contemporary real-estate industry (appraisal, finance, investment, taxation, and so on). Besides the most common calculations—of single payments and annuities, amortization schedules, monthly debt service, effective interest rate, and truth in lending—increasing use is being made of such statistical techniques as regression analysis and probability theory to help deal with uncertainty in capital investment analysis. Sensitivity analysis and risk analysis are becoming indispensable for evaluating future investment scenarios. Competent commercial software to assist in this process is emerging.

The widespread use of sophisticated mathematical analysis has been fostered by the use of computers. Indeed, many of the techniques in use today would not be possible without the processing power of modern computers. The accounting functions required for external financial statements are now routinely accomplished on computerized systems; and more specialized internal managerial functions for accounting and planning increasingly rely on computers. Integrated, comprehensive databases and computer systems are required to fuse all aspects of real-estate activity into a powerful tool for managing and planning facilities.

EIGHT

IMPLEMENTING A FACILITY MANAGEMENT SYSTEM

The most common question facility managers ask about switching to a computer-aided system is "when does it become economically practical to do so?" Again, a central theme of this book is pertinent: procedures, staff, and training are at least as important as (and possibly more important than) automation to the system's success. With that said, two basic factors are involved in the economy of using a computer program on a project: the amount of data (usually directly proportional to the employee population); and the change rate in the data (more closely related to the status of the organization than to its size).

Large projects (perhaps 400,000 square feet and greater) benefit significantly from automation and may be impossible to manage without data processing. Smaller projects cost less to manage using a computer-aided system, so an organization may be competitive in bidding in this range (100,000 to 400,000 square feet). And once a computer program and its related data collection procedures are installed in a facility management organization, projects of any size should be at least as well handled on the system as off the system.

To determine whether a computer-aided system will be cost-effective for a particular organization, the manager must get a handle on the organization's specific needs, with the aim of evaluating how well a particular product fits into the office process. Often the first step is to become much more explicit about that office process. How does the organization's facility management process work? What are the steps? Who is responsible for each step on a project?

CHOOSING A SYSTEM

Chapter 3 discussed in some detail the elements of a cost/benefit analysis for a facility management system. Figure 8-1 is a rather simple illustration of the selection process.

Acquisition Plan

The first step in choosing a system is to have an acquisition plan. In many respects, an acquisition plan is a business case that states, "This is what we want to do, and this is why it's economically sound for us to do it." The plan should not only be able to answer top management's questions of "What?" and "Why?" (in order to achieve funding approval), but also "When?" and "How?" (in order to successfully implement and use the system). At minimum, the acquisition plan should contain anticipated functions, costs, benefits, and return on investment (ROI).

The functions of the anticipated system—

Figure 8-1. Selecting a computer system. (Courtesy of Computer-Aided Design Group)

will be considerably longer (up to a hundred pages) and will include all the elements of a shorter plan as well as the following:

- A road map of the steps to be followed in acquiring a facility management system, the most important of which is building a solid ROI case by gathering appropriate cost/benefit data about the organization
- A "selling document" that shows top management the thought processes and business reasons behind the proposed acquisition

The acquisition plan generally should contain the organization's application priorities and planned staging (timing) of the applications to be acquired and implemented. This may be ordered by benefit-to-cost ratio with a "stop implementation" line drawn, as described in chapter 3. A discussion of funding sources and disbursement milestones also may be useful. Such a discussion may be brief, or it may be more extensive when funding is shared by many different cost centers, necessitating the examination of cost sharing allocation methods.

If the acquisition is being made or primarily implemented by people other than the ultimate users of the system, then identification of the key users along with provisions for "user representation" (to give them input in the process) should be part of the acquisition plan. Such would be the case when, for example, an organization's data processing group takes the lead in acquisition and implementation on behalf of the ultimate facility management users.

Vendors and consultants may assist in developing an acquisition plan. The International Facility Management Association (IFMA) is an excellent source of aid. IFMA's Computer Applications Council is in the process of creating and promulgating a "Draft CAFM RFP" (Computer-Aided Facility Management Request for Proposals) that promises to be a good starting point for plan development. The Computer Applications Council plans to make available to IFMA members much of the material presented here in chapter 9, along with other information.

exactly what it is to do—are central and must be stated in the acquisition plan. Ideally, functions should be articulated first in terms of the end result (output) and only secondarily in terms of the inputs and algorithms necessary to achieve the desired outputs.

Costs, benefits, and ROI, the fundamentals of any business case, are covered in chapter 3.

In organizations where there is broad consensus and funding for a facility management system, the acquisition plan will be a relatively short document (perhaps ten pages long) that will serve as a guide to the following steps:

- Consideration of specific applications, systems, and vendors
- Narrowing the selection and making contractual agreements
- Training staff and preparing data and procedures
- Implementing the system and establishing ongoing operations

In other organizations the acquisition plan

Specification

Next, a specification is needed. This should contain information on the following points:

- Specific functional tasks (what to store, what to process, what to answer, and so on—usually specified most succinctly in terms of the desired output)
- Computational environment (compatibility)
- Degree of comprehensiveness required
- Degree of integration required
- Benchmarks to be met

The specification of a facility management computer system is essentially a list of the requirements that the desired system should meet. The level and type of detail of such a list may vary; thus, specifications range from just a few pages of desired results to hundreds of pages describing technical operating environments, mean response times, formulas to be used, capabilities to represent particular types of problems, amount and performance specifications for training (for example, "At the end of training, the trainee shall be able to . . ."), and so on. The level of detail incorporated into the specification is determined by the acquirer, and it depends on many variables, including the following:

- Size and scope of the organization acquiring the system, including number and type of users
- Number of applications to be addressed
- Desires of management regarding level of detail, frequency of reporting, and so on
- Degree of technical sophistication, including algorithmic and procedural capabilities (for example, "The system should be able to predict growth of departments using the following compound formula . . .") as well as computer automation sophistication
- Whether the system is to be acquired "turnkey" (hardware and software) and "stand alone" (not to be interfaced with other systems and databases) or whether it must fit into existing automation hardware, software, or database environments (including an operating system and other applications)

A schematic illustration of the process of preparing a specification is shown in figure 8-2.

Caveat: "Make versus Buy"

In any new discipline, some managers will opt to design their own software. Indeed, facilities managers have only relatively recently had multiple serious alternatives to writing it themselves. Many vendors now understand the need for facility management systems, however, and competent off-the-shelf packages are available.

The advantages of custom software are apparent: closer fit and overall responsiveness to the organization. The disadvantages include significantly greater cost (especially for maintenance) and higher risk of obsolescence. Eileen Carstairs, writing in *Corporate Design and Realty* (March 1985), has this to say about "home-cooked" facility management software:

> The advantages of having a system specifically tailored to your own needs and developed by your own personnel are apparent.
>
> Problems can arise, however, in ensuring that the systems will be responsive to current and future advances in technology; often there is no ongoing effort to maintain an up-to-date data base and to add new enhancements to the system. Given the increased complexity of facility and property management and the lack of skilled manpower necessary to develop sophisticated programs, most small and medium-sized organizations no doubt will continue to rely on commercially prepared software and, only if necessary, modify it to suit their needs.

Even if an organization on its own can develop a highly suitable system, it will be hard pressed over time to support, maintain, and enhance the system as efficiently as would vendors whose only mission is to develop facility management software. With a custom system, the entire cost of maintenance must be borne by a single organization. Vendors are able to spread those same costs over hundreds of licenses, significantly reducing the costs to any one user.

Many years ago, CADD users faced the same choice, and those who developed their own systems were ultimately forced either to make costly conversions or to become unintended (and

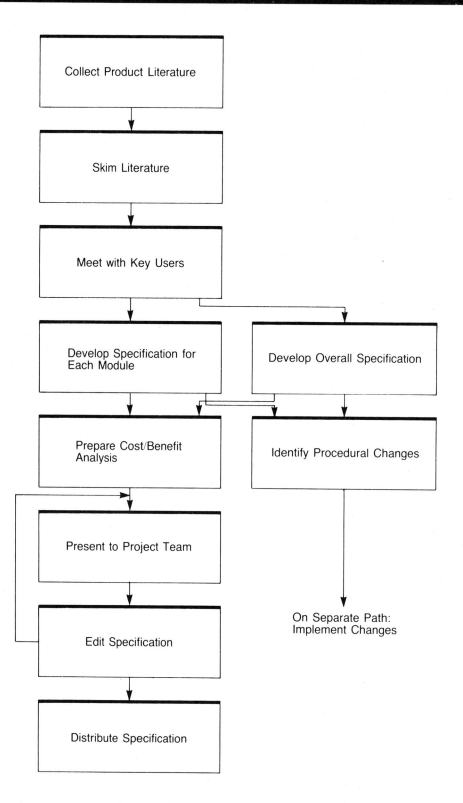

Figure 8-2. Preparing a system specification. (Courtesy of Computer-Aided Design Group)

sometimes unwilling) vendors themselves (in order to expand their installed base to cover enhancement costs). Many of the early developers of CADD software were design firms in the AEC (Architecture, Engineering, and Construction) business. A few were developed or sponsored by professional societies, the government, or other large end-users. Virtually none of those systems is commercially viable today. In fact, most of the design firms that entered the business with the publicly stated intent of becoming (and remaining) vendors have had less-than-impressive staying power—perhaps because software development, sale, and support represent a fundamentally different business from those that sponsored or undertook them. The few commercial success stories that do exist are mostly cases in which the initial developer formed a partnership with a computer or software firm that understood the commercial software business.

We live in an age of increasing economic specialization. The technically robust and economically viable software systems of today (in CADD and in many other disciplines) are produced and supported by organizations specializing in that business. History is repeating itself today in the field of facility management systems. Systems are being developed today not only by AEC firms but by major space owners, users, and corporations. Those that will be in use a decade from now are likely to be those marketed, supported, and maintained by large, viable computer, software, and consulting organizations.

Software Checklist

Once the list of potential software systems has been narrowed down to those best suited to the specification, the following final checklist can be applied to the set of final choices:

- Is the system comprehensive? Does it address all areas of facility management?
- Is it expandable and modular? Can functions be added to it in the future as needed?

- Is it available free of hardware bundling (a requirement of the vendor that both hardware and software be purchased together), and can it be supported in an industry-standard or widespread computational environment?
- Is it compatible with existing CADD systems and the organization's primary (mainframe) and departmental (minicomputer/office automation) hardware and software? Can it coexist and share data with other related computer applications (personnel, fixed assets, strategic planning, accounting, business records, and other centralized applications)?
- Are comprehensive initial training and (perhaps more important) solid ongoing means of educating new staff already in place?
- Is the system really "user friendly"? Can the organization's staff learn how to use it well? Is the interface consistent in all functions and applications? What are the quality, quantity, and accessibility of the training?
- Does it include extensive procedures? (Software alone will not instruct staff how to do a furniture inventory, for example. A true system must include procedures, whether vendor-supplied or developed in-house.)
- Does it include an industry-standard database management system? (A good DBMS allows quick canned reports to be produced, by means of efficient network queries and specialized, fully relational reporting in SQL [Structured Query Language].)
- Is the software robust and designed for a large organization? (Issues of concurrency and security must be thoroughly addressed. Hardware failures and unintentional user errors occur regularly in any environment. A database shared by many users must be capable of concurrent access and must have strong backup, journaling, audit trail, batch input, and administrative functions.)

Prioritization

The best facility management software systems are modular and can be phased in over time. A professional consultant or software development vendor works with clients to help determine the priority according to which modules should be acquired and phased in. One frequently used method is to evaluate each module on the basis of productivity, availability, cost, cost/benefit relationship, frequency of use, and level of technology. Numerical values are assigned to each criterion, and the modules are ranked according to the total points accumulated by each.

PHASED IMPLEMENTATION

A program of phased implementation is desirable for a variety of reasons, including spreading out the cost and the degree of assimilation of the system required in order to operate it. Often a two- or three-year time frame is necessary (assuming the availability of funding) to "round out" a full program. Vendors and professional organizations can be helpful in developing specifications based on what is realistic, practical, economic, and available. Figure 8-3 illustrates a sample implementation schedule for one module of a facility management system. The process involved is described in the ensuing subsections.

Developing a System Overview Specification

A system overview specification is the first step in specifying the facility management system. It should be developed to identify types of data input and data interaction (both initial and future), the size of fields, estimated storage capacities, and other important operating characteristics of the total system. The final specification includes all facility management system (FMS) components and prescribes such things as the following:

- The environment in which the program(s) should run
- The types of input, processing, and output devices available for utilization
- The content and type of operating and user manuals to be provided
- The controls and safeguards that must be incorporated to avoid tampering or inadvertent unauthorized database changes
- The amount of information that must be stored in each category of the database

- The fields necessary and the formats desirable for displaying information

The system overview specification should also provide diagrams that clearly indicate the flow of information from one FMS module to another, as well as the overall operational linkages of the system.

Identifying Linkages

As has already been stated, data entry is the largest single cost of any computer application. Eliminating potentially redundant data entry by "bridging" to "peer databases" within the organization yields large payoffs in reduced data entry time, cost, effort, and error. For these reasons, it is important to identify (where known) the sources of data, the calculations that are to be used in each functional area, and the ultimate destination of information developed or processed in each module. The activities of modules that will be implemented later should be supported, and specifications regarding informational requirements in each module must be identified. Care taken at this step can yield large benefits later. The ease and economy of data entry and verification together with the reduced redundancy that results from bridging to peer databases can provide the single largest economic benefit of a facility management system.

Selecting Data Formats

It is desirable for the user to develop in rough form the required data display and report formats. Of particular interest are the potential length and precision of each piece of data, the number of data elements that may need to be processed or displayed on summary reports and charts, and the selection of a format for the output of the data that lends itself well to the desired uses. A good place to begin the analysis is with the format of the present, manually generated information. The requests of management and other users should then be in-

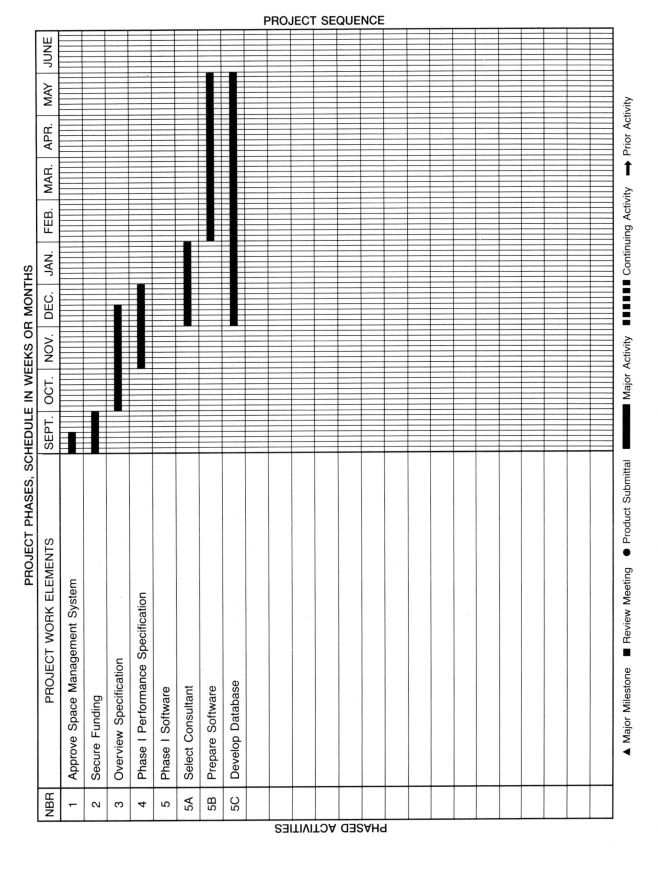

Figure 8-3. Implementation schedule. (Courtesy of Steinmann, Grayson, Smylie)

corporated into the desired report and presentation formats. Examples of report and presentation formats that may be analyzed include rent allocations, space utilization and occupancy profiles, future requirements projections, adjacency information, building stacking profiles, statistical analysis reports (relating to long-range planning, analysis of alternatives, and database management), and so on.

Probably the most important decisions in this area relate to the formats chosen for top management summaries and decision-making aids.

Preparing Performance Specifications

For each of the FMS components planned for first-phase implementation, a performance specification should be developed, outlining desired/required response times, personnel resources, training results, machine resources, and so on.

The performance specification also may perform the following functions:

- It may elaborate on the requirements identified in the overview specification.
- It may provide samples of data input and output formats.
- It may quantify the magnitude of data that must be stored and analyzed by the module.
- It may discuss operating systems.
- It may itemize necessary controls and safeguards.
- It may dictate the hardware to be utilized.
- It may prescribe programming languages available for interface or the language to be used (if the organization is building all or part of its own system).
- It may specify existing databases that the system must be able to interface with and perhaps interface standards or formats (if inflexible) as well.
- It may identify documentation that must be provided by the FMS system vendor.

Requesting Proposals

Once the performance specification is available, a request for proposals (RFP) can be developed and forwarded to qualified firms. An RFP may be as simple as asking vendors to submit their standard proposal of product, function, terms and conditions, maintenance and warranties, and so forth, or it may be quite complex. Chapter 9 contains vendor qualification information and questions that may be used both to develop an RFP and to custom tailor it to the needs of a particular organization. Chapter 9 also contains a criteria checklist that may be used to evaluate the responses. As mentioned earlier, the Computer Applications Council of the International Facility Management Association (IFMA) is in the process of developing a standard RFP for CAFM (Computer-Aided Facility Management) utilizing this information.

Lists of current potential suppliers (and periodicals in which such listings are continually updated) may be found in the appendixes.

Providing Training

Time must be allocated and a budget must be provided for training facility management personnel in the use of the FMS—both initially and on a continuing basis. A serious commitment on the part of the organization to set aside this required time is one of the most important variables affecting success.

Assembling the Database

Utilizing forms and procedures developed prior to installation/implementation (and in concert with procedural documentation supplied by the successful vendor), personnel should begin to gather current and historical information, to code prepared data input forms for loading into the FMS, and to assemble a comprehensive database. The primary focus should be on assembling the resources inventory: space (current, surplus, or available), furniture and equipment, space utilization, leased-space occupancy profiles, costs and their allocation, personnel, and so on. The contents of the resources inventory is discussed in chapters 4 and 5.

Developing the Facility Management Organization

Strengthening the space management organization and supplementing it with additional staff resources (if required) merit attention in the process of phased implementation. This is discussed in the following section.

FACILITY MANAGEMENT ORGANIZATION

To expedite implementation of the facility management system, it is sometimes desirable to make organizational changes—one of which would be to expand the responsibilities of the facility management unit (providing additional staff resources, if needed).

The facility management organization should include the following three components if it is to function properly, maximizing space utilization efficiency and cost effectiveness:

· A clear delineation of responsibilities, duties, and procedures to follow
· The authority necessary to execute those responsibilities properly in the years to come
· Staffing personnel trained and experienced in discharging their duties

The subsections that follow discuss each of these three areas.

Responsibilities

An effective space management organization assembles professionals from a number of different organizational units into a unit that concentrates on issues central to developing an effective facility management program. The space management unit should have equal status with (or should incorporate) any units whose existing functions include real-estate appraisal, leasing, property acquisition, and maintenance.

Eventually, the space management organization, because of its broad, long-range perspective, may become the chief directional force in an already effective real-estate-related organization.

Regardless of how they are assembled, the personnel working in the space management unit should be assigned such responsibilities and duties as leasing space, acquiring property, and day-to-day maintaining and operating the existing space inventory. In all cases, they should be provided with sufficient staff to execute those duties properly.

Jim Steinmann suggests that the following responsibilities and duties focusing on the space management process should be added to the organization's existing responsibilities:

· Conduct studies of space needs, and develop and monitor plans regarding long-range space needs.
· Manage space, and monitor its utilization.
· Prepare and review schematic and development drawings reflecting interior spaces, to ensure that space standards are maintained.
· Schedule and coordinate implementation of facility development processes.
· Conduct building feasibility studies.
· Assist other departments with their future-growth projections.
· Conduct necessary negotiations for executing leases and disposing of expendable surplus property.
· Develop and prepare an annual update of future personnel and space projections for all departments and space in the inventory.
· Prepare and consolidate annually a capital improvements program confined to the building environment that supports all personnel, equipment, and special areas.
· Maintain up-to-date record documents to reflect all existing leases, owned spaces, area assignments, area factors, personnel assigned to each organizational unit, furniture inventories, and other fixed-asset control procedures related to the management and utilization of space.
· Prepare an annual space management report that documents existing conditions, projects future requirements, and updates the existing facility acquisition and space development plan.
· Monitor the utilization of space, review department

applications for additional or alternative space, negotiate requirements, approve final space allocation distributions, and prioritize requests for space alterations, relocations, or expansions.

- Prepare and administer a detailed procedure for calculating future space requirements, monitor the allocation of space and ensure that all space actions (new construction, remodeling, or leasing) are completed in a manner consistent with the requirements identified in specific space-programming data reports.
- Analyze alternative furniture systems, and develop a procedure (not necessarily based on minimum cost) for developing furniture system standards and for acquiring furniture systems to be employed in all future space development or remodeling activities.
- Maintain continuous awareness of new developments in space utilization and systems furniture, from time to time soliciting additional bids and proposals for new items; make final selections of equipment and furniture to be employed in interior environments; maintain standards for their distribution, their utilization, and the quality of plans and designs prepared by staff and consultants.
- Refine and administer a professional selection system to provide optimum-quality professional consultation on space programming, long-range planning, master planning, space planning, and interior and architectural design for space in all owned or leased facilities; develop selection, review, evaluation, and contractual procedures for these purposes.
- Monitor the scope of work, the responsibilities, the schedule and cost performance, and the quality of services provided by consultants on specific projects to be implemented. (In this regard, the space management unit becomes a project management organization, with personnel assigned to monitor each design, construction, or remodeling project.)
- Review and approve interior design and space plans developed by others; monitor the solicitation of proposals from and the selection of vendors for equipment and furniture; assist in negotiating annual procurement contracts for appropriate furniture and equipment items, to ensure adherence to standards and to optimize space utilization.

The above list of responsibilities and duties tends to focus on office-space environments, but the tasks appropriate to other types of occupancies are similar.

Authority

Along with these responsibilities must come the necessary authority to carry out the delegated assignments effectively. The space management organization must have clearly defined authority with regard to the technical matters it will be directing and administering. In addition, the unit's authority must not be subject to undue influence or change by others (including management), if the space management system is to operate in a timely and cost-effective manner. In general, the space management organization must be able to make independent decisions (within management-defined parameters) about the following matters:

- The amount of space assigned to each department
- The layout and spatial organization of the space
- The geographic location of new buildings and lease space
- The terms and conditions under which those spaces will be developed and provided to the departments

These decisions are economic and technical, and (within the boundaries of the criteria and economics dictated by top management) they should not be subject to influence by management from other departments.

More specifically, authority to perform the following functions should be vested in the space management organization:

- Development and monitoring of space, environmental, and furniture standards
- Review and selection of furniture and equipment items that are part of the space standards system
- Development of space requirements, and specification of space to be allocated to various organizational units
- Recommendation of consultants or architects to be awarded a consulting assignment for each project, not based solely on a low bid (additional criteria should result from competitive negotiation after

a definition of the project and a detailed scope of work have been established)

- Establishment and monitoring of professional selection procedures and formats, including development, review, negotiation, and administration of related contracts
- Development of prearchitectural building programs and other letters of instruction that might be provided to the professional consulting community to assist in the proposal stages and to direct initial design activities
- Determination of space assignments and area blockouts for all departments included in a specific building
- Review of alternatives, development of final documents, and rearrangement and relocation of departments within general existing geographical areas
- Review and disposition of requests by departments for additional space, space adjustments, construction projects or relocation to alternative quarters
- Determination of whether departments should utilize existing furniture or be provided with new furniture, and (if the latter) whether that furniture should be systems furniture, open-office planning, or private office environments
- Approval or rejection of any request to purchase furniture or equipment that consumes floor space—of any type, at any time, and in any quantity, by any employee or department (consistent with the organization's strategic or business plan)

A number of important decisions must be made annually by the hierarchy of authority. These include the following:

- Final decision as to the primary occupants of any single department or decentralized facility to be constructed
- Decisions as to whether a facility should be leased or new construction should be recommended and included in the budget process
- Final review of space acquisition, remodeling, and space management activities that are to be included in the annual budget
- Final review and approval of long-range master plans for facility development
- Prioritization of the annual work program to be completed by the space management organization

Other decisions are appropriately made only at the highest possible level of authority:

- Funding for implementation of the selected components of the facility development and acquisition strategy
- Final decision as to funding for construction of a building to house a particular department or a combined facility
- Final review and adoption of a long-range facility master plan

Other than in these instances, department executives and other high-level management personnel generally should not participate in implementation decisions—particularly in decisions regarding the amount of space to be provided, the quality of that space, the type of interior planning to be utilized, the furniture to be employed, the location of agencies within buildings, or the identities of particular properties or facilities to be procured or leased. (Top management should either trust its facilities staff or replace it.)

Staffing

The director of a facility management unit can be both landlord and "space czar," with ultimate responsibility and authority for projecting needs, providing space, monitoring space utilization, and directing related maintenance and operational activities. The status of the space management unit should be equal to that of the property acquisition, disposal, and leasing divisions.

Facility management groups generally are headed by a full-time director who is assisted by a full-time secretary. One frequent form of organization for a facility management unit consists of division of the unit into three subunits—long-range planning, requirements programming and analysis, and space planning and interior design—although at inception (for the first year or two) certain subunits are sometimes consolidated.

Long-range Planning: The head of long-range planning is responsible for overseeing annual facility development plans, for monitoring and updating the long-range facility utilization program, for preparing annual reports to management, and for maintaining inventory and record data on space utilization.

Requirements Programming and Analysis: The head of this subunit is responsible for developing programming procedures, for completing certain technical projects, for monitoring requests for space from the various occupant groups, and for directing the needs programming and analysis of alternative space/facility development strategies. This last area includes such duties as life-cycle cost analysis, development of prearchitectural documents, and completion of various more precisely focused space programming studies. Requests for new space pass initially through the head of this group, who works closely with the long-range planning group to determine priorities and procedures necessary to implement approved requests. The head of the requirements programming and analysis group ultimately will require the support of one or two technical aides.

Space Planning and Interior Design: The head of the space planning and interior design group should have a background in space planning and interior design and probably will require the support of personnel with backgrounds in interior design, construction/implementation document development, space planning, drafting, and graphics services. A support (production) staff usually is essential for this subunit when it becomes fully activated.

A space management organization usually includes associated clerical support. Automation also brings the need for database administration and for an individual with title, responsibility, and authority to maintain the integrity of machine-readable information by implementing security, backup, and similar procedures. *Facilities Design and Management* magazine found that approximately 1 percent of the employee population of large corporations and institutions was involved with facility management (broadly defined to include facility engineering, planning, real-estate management, and the like).

NINE

CAVEAT EMPTOR

For readers who are seriously considering acquiring an automated system for facility decision support, a few guidelines are in order. Three questions are particularly relevant:

1. How do you decide whether computer-assisted facility management tools are appropriate for your organization?
2. How do you decide whether they are cost-effective?
3. If they are cost-effective, how do you choose which system to apply?

The first two questions are best handled by means of dispassionate and comprehensive cost/benefit analysis, as discussed in chapter 3. Several excellent books and articles in the trade publications have been written on this particular subject. A good first step might be to attend a meeting of the International Facility Management Association (IFMA) and to take one of their courses. Consultants can be of great benefit here, too, but only after sufficient specific and relevant experience has carefully been verified and conflicts of interest have been eliminated.

The third question is the subject of this chapter. What specific questions should be asked of a facility management system vendor? The selection criteria checklist that follows includes most questions of general importance. Topics of unique or special interest to your organization should be added to the list. The checklist is not a substitute for "road testing," a thorough independent investigation, and other appropriate steps.

Some vendors of facility management systems are primarily or exclusively CADD vendors. Some specialize in CADD and make only a token effort in the area of facility management; others are primarily or exclusively facility management vendors and may not even offer CADD. The checklist is designed to evaluate vendors on the basis of facility management (not CADD) criteria only, and it should not be applied to CADD systems in an attempt to evaluate them.

The selection criteria checklist is presented without any scoring or weighting of the items in it. Such weighting should be decided on the basis of which items, if any, are more important than others for your organization.

FACILITY MANAGEMENT SYSTEM SELECTION CRITERIA CHECKLIST

Vendor History

This section deals with the general reputation, consistency, and strength of the vendor in the industry.

1. **Reputation, standing:** What does the "street" think of this firm? Do people say good things about their honesty, integrity, and fairness? Is the firm well represented and well thought of by trade groups, professional societies, and so on? Is it adequately represented in the literature?

2. **Size of installed base:** What size is the installed base? The number of installations a firm has (if large) is a good indicator that the firm will be here tomorrow.

3. **Quality of installed base:** Are the organizations in the vendor's client list similar to yours? Are they "name" organizations or unknowns?

4. **Support:** What steps does the vendor take to support its products—toll-free hotlines? updates? (how often?) anything else? Is there a priority-escalation policy for unsolved problems? How accessible and competent are the support staff? How accessible is management?

5. **References:** Is the vendor willing to give you many references, without prodding? When you call the references, are they happy? Would they make the same choice again? Do they seem to know what they are doing? Are they solving problems similar to the ones you want to solve?

6. **Consistency, tenure of management:** How long has the vendor's present top management team been at the helm? A consistent management team helps ensure that the company is stable and takes the long view. It also indicates that the vendor is less likely to experience swings in policy or to emphasize different portions of its business in the future.

General System Characteristics

This section discusses general attributes (ones that are not application-specific) of vendors' systems—things that affect overall performance and suitability regardless of the project (whether, say, a lease analysis or a stacking plan is envisaged).

1. **Industry-standard computing environments:** Does the vendor's system work in an industry-standard operating environment? This means on an unmodified computer from a well-established major vendor. Organizations that utilize such environments for facility management and other applications are able to take advantage of advances in hardware without disrupting applications and can run programs from different vendors concurrently, thereby making maximum use of computer resources. Nonstandard hardware or operating system software bears the risk of becoming an "orphaned" hardware product because the vendor goes out of business, changes direction, or cannot keep up with the industry's continuing price and performance standards. The computer industry is consolidating, moving away from vendor-specific products to standardized ones that can talk to each other. At present, the number one and number two computer vendors in the United States are International Business Machines Corporation (IBM) and Digital Equipment Corporation (DEC). Their operating systems are overwhelmingly prevalent in offices of major facility management user organizations.

2. **Availability without hardware:** Are the vendor's facility management software products available unbundled (without hardware)? Such products do not force their users to buy entire "turnkey" or "bundled" systems; instead, they allow their users to obtain the application software from the vendor of their choice at the time of their choice, and to install it on the equipment of their choice. As operating environments become more and more similar and as software becomes more and more portable, the advantages of selecting software without hardware will increase further.

3. **Price:** What is the cost of the system, fully configured, with all of the required hardware options and software features?

4. **Data entry:** Does the system use a single, integral, industry-standard database management system? Because the cost of clean data entry is the largest single part of any computer application, information should be entered into the system only once. Moreover, because many users (often in different departments) need access to the same information, individual users should be able to download desired portions of the database onto personal computers or workstations for their own analysis while the master database remains integral.

5. **Standard reports:** Are "canned" standard reports available to allow users to begin successful use

of the decision-support system quickly? What quantity and variety of such reports are offered?

6. **Color business graphics:** What types of color business graphics are offered? Are "canned" standard graphics available to allow rapid start-up use?

7. **Industry-standard query language:** Is an ad hoc query capability available for special or one-time information needs? In this area, SQL (the language, not necessarily a particular program product) is already the standard.

8. **Report writer:** When a continuing need arises for a report that is not part of the standard menu of "canned" formats, does the system allow the user to create and "can" a new one, using a report-writer language? How powerful is this language? Will it permit creation of reports in the exact format users desire?

9. **Unloading to PC:** Can individual users download portions of the database to their own personal computers or workstations for analysis in a spreadsheet, incorporation in a word-processed report, or some other purpose? Can this be done easily? What formats (such as the industry spreadsheet standard, DIF) are available?

10. **Interface to other databases, systems:** How easily can the product exchange information with other databases in the organization, such as those for personnel, strategic planning, accounting, architecture, interior design, CADD, and fixed assets? Of the features that reduce potentially high data entry costs, interface to other databases is perhaps the most important.

11. **Documentation quality:** Is the accompanying documentation comprehensive, easy to use, illustrated, highlighted, well-indexed, logically organized, current, and error-free?

12. **Documentation quantity:** Are all aspects of the system documented in detail, including such esoteric but critical functions as backup, error recovery, and database administration?

13. **Training:** Is training adequate? Is it sufficiently lengthy? Does the price include enough training for enough people? How is retraining handled for both existing staff and new staff? Does the training include operation of hardware, support and administration, and surrounding procedures?

14. **Computer-aided instruction:** Does the vendor include software for computer-based training,

to ensure that new hires can be brought up to speed quickly?

15. **Procedures:** How would you characterize the amount and quality of the procedures that the vendor and its products include? What is the degree of fit they will have with existing procedures within your organization? Whether or not an organization decides to automate, procedures (for accurate and timely data capture, use of the information when processed, and so on) can be the single most important factor in the success of a facility management system. It is unlikely that even the best vendor-produced procedures will fit perfectly into a given organization without modification. How universally applicable and how flexible (modifiable) are the procedures? Is the vendor willing to help?

16. **Unified user interface:** Are the many different facility management functions presented to the user in a consistent way, so that learning is easier and so that infrequent or casual users can experience a high degree of success?

17. **Mode of interaction:** What mode of interaction between user and computer does the system incorporate? Is it appropriate to your organization? Screen-form menus are the easiest method of interaction for naive, new, frightened, or infrequent users, although they are frustrating for expert users (*problem-oriented* [command] *languages* are best for them). In any case, the overwhelming majority of facility management system users do not aspire to become computer jocks.

18. **Major CADD system interfaces:** Can the system interface and share data with the CADD system in use within your organization? Can it do so with the industry market-share leaders among CADD systems (which one day may be in use within your organization or its subsidiaries, partners, or affiliates)?

19. **Ease of use:** Subjectively, how easy is the system to use? During your "road test" of the system, can your people accomplish their desired tasks?

20. **Ease of learning:** How quickly do your people become familiar with the feel and flow of the system?

21. **On-line help:** Does the system allow the user at any time to ask for help in a consistent fashion (either by pressing a special help key or by typing *help*, a question mark, or some other alphan-

umeric key)? Is the help supplied context-dependent and consistent with the printed documentation?

22. **Robustness:** During your "road test" of the system, how did it respond to unintentional (and intentional) errors? Could you make the system "crash"?

Specific Applications

This section identifies many specific functional areas and broad classes of problem-solving needs (called *applications*) within the discipline of facility management. Although the applications listed are ones that should be considered by most organizations, it would be folly to try to implement all (or more than a few) at once. Nonetheless, their existence and general quality should be investigated, since you may be adding many of them later on. The capacity for communication among these applications is an important characteristic in evaluating them. These applications are discussed in more detail in chapters 4 and 5.

1. **Inventory:** This is a resources inventory that classifies spaces, furniture, equipment, activities personnel, leases, cost data, and so on, by organization, location, or other appropriate criteria.

2. **Requirements programming:** This is a needs inventory that classifies data by the same criteria as are used in the resources inventory.

3. **Adjacencies and relationships:** These consist of the material and personnel/customer flow and the communications needs of the organization.

4. **Optimized location planning:** This type of planning includes automated production of optimized region, city, campus, site, and wing allocations.

5. **Optimized stacking planning:** This type of planning covers the automated production of optimized allocations to floors of multistory structures.

6. **Optimized block planning:** This type of planning involves the automated production of optimized allocations (block diagrams) on floors.

7. **Drafting coordination:** This involves either integration of CADD (as in the case of a CADD vendor) or close coupling to a CADD system (or better yet, multiple CADD systems), for the easy interchange of data.

8. **Master planning:** This type of planning provides executives with high-level "what if?" decision-making ability for maintaining many alternative scenarios and for exploring cost, time, and staging implications of the various scenarios.

9. **Cost accounting:** This involves aggregation and allocation of detailed cost information.

10. **Real-estate/financial strategy:** This application assesses lease versus build versus buy, present-value life-cycle costs, and so on.

11. **Purchasing coordination:** This involves coordination with furniture and equipment inventories, specifications, catalogs, purchase orders, and the like.

12. **Group administration, budgeting:** This application covers facility management office tools and personnel administration tools.

13. **Project, construction management:** This includes CPMs, PERT charts, ticklers, line item status checkers, and so on.

14. **Maintenance planning:** This type of planning covers planned maintenance reminders, unplanned maintenance analysis and recommendations, and the like.

15. **Site management:** This involves management of between-building infrastructure, services, and relationships.

16. **Design analysis:** This application relates to analysis, critique, code compliance issues, and other aspects of a project's design.

APPENDIXES

The information in these appendixes unfortunately becomes out of date *very* rapidly. (It is to be expected that some of the resources listed will have changed even before the initial printing.) Nevertheless, in an emerging discipline such as facility management, resource lists are much needed.

APPENDIX A

EDUCATION

UNIVERSITY PROGRAMS

University-level (and even graduate) programs specifically dealing with the professional discipline of facility management are growing rapidly in popularity. Only a few of the current programs are listed here.

Canadian School of Management
820 Renaissance Plaza
150 Bloor Street
W. Toronto, Ontario M5S 2X9, CANADA
Margaret Ebrecht, Representative

Carnegie Mellon University
College of Fine Arts
Department of Architecture
Schenley Park
Pittsburgh, PA 15213
Professor Ulrich Flemming

Cornell University
Programs in Professional Education
B12 Ives Hall
Ithaca, NY 14853
Professor Frank Becker

Grand Valley State College
Facilities Management Program
Allendale, MI 49460
Robert D. Vrancken, Director

Massachusetts Institute of Technology
Office of Facilities Management Systems
77 Massachusetts Avenue, E19-451
Cambridge, MA 02139
Keron Cyros, Director

Michigan State University
College of Human Ecology
Department of Human Environment & Design
204 Human Ecology Building
East Lansing, MI 48824
Professor Jane Stolper

Northeastern University
Building Technology Program
370 Common Street
Dedham, MA 02026
Christopher Cassidy, Representative

University of California at Irvine
Program in Social Ecology
Irvine, CA 92717
John West, Director

Wentworth Institute of Technology
550 Huntington Avenue
Boston, MA 02115
Frederick Gould, Representative

PRODUCTS

Educational products also are beginning to appear. The International Facility Management Association (IFMA) teaches a course entitled Principles of Facility Management; attendees of this course receive a manual entitled *An Overview of Facility Management,* which includes an excellent glossary of facility management terms. It also contains IBM-PC computer-aided instruction diskettes corresponding to the contents of the manual. This product is produced by the Computer-Aided Design Group and is licensed to IFMA course attendees as part of the course. It also is available separately from the following address.

An Overview of Facility Management
Computer-Aided Design Group
4215 Glencoe Avenue
Marina del Rey, CA 90292

COURSES

American Management Association (AMA)
135 West 50th Street
New York, NY 10020
(Various office-space planning courses are offered.)

The Computer-Aided Design Group
4215 Glencoe Avenue
Marina del Rey, CA 90292
(Various seminar courses are offered, including the following two: Overview of Facility Management, and Identifying and Estimating the Benefits of Facility Management Systems.)

Cornell University
Programs in Professional Education
B12 Ives Hall
Ithaca, NY 14853
(Courses include Tools and Techniques for Facility Planners and Managers Seminar.)

International Facility Management Association (IFMA)
Summit Tower, Suite 1410
11 Greenway Plaza
Houston, TX 77046
(Various courses are offered, including the following two: Principles of Facility Management, and Facility Management Automation.)

Massachusetts Institute of Technology
Office of Facilities Management Systems
77 Massachusetts Avenue, E19-451
Cambridge, MA 02139
(Various courses are offered.)

New York University
School of Continuing Education
Seminars in Building Construction
11 West 42nd Street
New York, NY 10036

Steinmann, Grayson, Smylie
6310 San Vicente Boulevard
Suite 550
Los Angeles, CA 90048
(A course is offered in Facility Utilization Audits.)

University of California at Irvine
Program in Social Ecology
Irvine, CA 92717
(Various courses are offered, including Introduction to Facilities Design and Management.)

APPENDIX B

RELEVANT PERIODICALS

FACILITY MANAGEMENT EMPHASIS

The following periodicals often contain significant facility management content.

A-E-C Automation Newsletter
7209 Wisteria Way
Carlsbad, CA 92009

A/E Systems Report
Design & Systems Research, Inc.
186 Alewife Brook Parkway
Cambridge, MA 02138

AM/FM Chronicle
Automation Group, Inc.
1708 East Kensington Road
Arlington Heights, IL 60004

BRB Research Briefs
National Research Council
2101 Constitution Avenue, NW
Washington, DC 20418

Building Economics
1221 Avenue of the Americas
New York, NY 10020

Building Operating Management
2100 West Florist Avenue
Milwaukee, WI 53209-3799

Building Sciences
National Institute of Building Sciences
1015 15th Street, NW
Suite 700
Washington, DC 20005

Buildings: The Facilities Construction & Management Magazine
427 Sixth Avenue, SE
P.O. Box 1888
Cedar Rapids, IA 52406

Business Facilities
121 Monmouth Street
P.O. Box 2060
Red Bank, NJ 07701

Corporate Design
Cahners Plaza
1350 East Touhy Avenue
P.O. Box 5080
Des Plaines, IL 60017-5080

Emerging Trends in Real Estate 1986
Real Estate Research Corporation (RERC)
72 West Adams Street
Chicago, IL 60603

Facilities
Bulstrode Press
8-9 Bulstrode Place, Marylebone Lane
London W1M 5FW ENGLAND

Facilities Design & Management
1515 Broadway
New York, NY 10036

Facilities Planning News
Tradeline
115 Orinda Way
Orinda, CA 94563

FM Automation Newsletter
9501 West Devon Avenue
Suite 203
P.O. Box 507
Rosemont, IL 60018

IFMA News
International Facility Management
  Association
Summit Tower, Suite 1410
11 Greenway Plaza
Houston, TX 77046

*Industrial Development Research Council
  Newsletter*
1954 Airport Road, NE
Atlanta, GA 30341

Industrial Engineering
Institute of Industrial Engineers
25 Technology Park
Norcross, GA 30092

INSITE on Facilities Management
Office of Facilities Management Systems
Massachusetts Institute of Technology
77 Massachusetts Avenue, E19-451
Cambridge, MA 02139

Journal of Property Management
430 North Michigan Avenue
Chicago, IL 60611

NACORE News
International Association of Corporate Real
  Estate Executives
471 Spencer Drive, S
Suite 8
West Palm Beach, FL 33409

NAIOP News
National Association of Industrial and Office
  Parks
1215 Jefferson Davis Highway
Suite 100
Arlington, VA 22202

SUPPORTING INTEREST

The following periodicals cover central issues
of facility management but also other areas of
related interest.

Administrative Management
1123 Broadway
New York, NY 10010

Architectural & Engineering Systems Magazine
3307 South College Street
Suite 337
Fort Collins, CO 80525

Architectural Record
1221 Avenue of the Americas
New York, NY 10020

Architectural Technology
1735 New York Avenue, NE
Washington, DC 20006
or c/o AIA Service Corporation
1735 New York Avenue, NW
Washington, DC 20006

Architecture: The Journal of the American Institute of Architects
1735 New York Avenue, NW
Washington, DC 20006

Area Development
525 Northern Boulevard
Great Neck, NY 11021

BRB
Building Research Board
2101 Constitution Avenue, NW
Washington, DC 20418

Building Design & Construction
1350 East Toughy Avenue
P.O. Box 5080
Des Plaines, IL 60018

Buildings Design Journal
6255 Barfield Road
Atlanta, GA 30328

CAD/CAM Alert
824 Boylston Street
Chestnut Hill, MA 02167

CAM-I News Alert
611 Ryan Plaza Drive
Suite 1107
Arlington, TX 76011

Canada's Contact Magazine
777 Keele Street
Unit 8
Concord, Ontario, L4K 1Y7, CANADA

CIM Technology
One SME Drive
P.O. Box 930
Dearborn, MI 48121

Computer Business
P.O. Box 45923
Los Angeles, CA 90045

Computer Communications Decisions
10 Mulholland Drive
Hasbrouck Heights, NJ 07604

Computer Graphics
50 West 23rd Street
New York, NY 10010

Computer Graphics Today
National Computer Graphics Association
2722 Merrilee Drive
Fairfax, VA 22031

Computer Graphics World
119 Russell Street
Littleton, MA 01460

Computers for Design & Construction
310 East 44th Street
New York, NY 10017

Computer Systems News
600 Community Drive
Manhasset, NY 11030

Computerworld
375 Cochituate Road
P.O. Box 9170
Framingham, MA 01701

Construction
7297 Robert E. Lee Highway
Falls Church, VA 22042

Construction News
P.O. Box 2421
Little Rock, AR 72203

Construction Specifier
601 Madison Street
Alexandria, VA 22314

Contract
1515 Broadway
New York, NY 10036

Datamation
875 Third Avenue
New York, NY 10022

Engineering News-Record
1221 Avenue of the Americas
New York, NY 10020

Expansion Management
514 10th Street, NW
Suite 800
Washington, DC 20004

Facilities Energy Report
P.O. Box 21129
Washington, DC 20009

ICP Business Software Review
9000 Keystone Crossing
P.O. Box 40946
Indianapolis, IN 46240

Industrial Design
330 West 42nd Street
New York, NY 10036

Industrial Development
1954 Airport Road, NE
Atlanta, GA 30341

Industrial Management
277 Lakeshore Road, E, #209
Oakville, Ontario, L6H 2R9 CANADA

Information Management
101 Crossways Park, W
Woodbury, NY 11797

Information Systems News
111 East Short Road
Manhasset, NY 11030

Information Week
600 Community Drive
Manhasset, NY 11030

Infosystems
Hitchcock Building
Wheaton, IL 60188

Interiors
1515 Broadway
New York, NY 10036

Manage
2210 Arbor Boulevard
Dayton, OH 45439

Management World
Administrative Management Society
4622 Street Road
Trevose, PA 19047

MIS Week
7 East 12th Street
New York, NY 10003

Modern Office Technology Magazine
1100 Superior Avenue
Cleveland, OH 44114

National Productivity Review
22 West 21st Street
10th Floor
New York, NY 10010-6904

The Office
1600 Summer Street
Stamford, CT 06904

Office Systems Ergonomics Report
3029 Wilshire Boulevard
Santa Monica, CA 90403

Office Technology and People
Elsevier Science Publishers
P.O. Box 211-1000 AE
Amsterdam, NETHERLANDS

Professional Office Design
111 Eighth Avenue
Suite 900
New York, NY 10011

Progressive Architecture
600 Summer Street
P.O. Box 1361
Stamford, CT 06904

Real Estate Business
430 North Michigan Avenue, #500
Chicago, IL 60611

Software News
1900 West Part Drive
Westborough, MA 01581

Urban Land
1090 Vermont Avenue, NW
Washington, DC 20005

The World of Work Report
700 White Plains Road
Scarsdale, NY 10583

APPENDIX C

PROFESSIONAL ORGANIZATIONS, SHOWS, AND EXPOSITIONS

PROFESSIONAL ORGANIZATIONS

ACD
Association for Computers in Design
1341 Ocean Avenue
Suite 236
Santa Monica, CA 90401

ACM
Association for Computing Machinery
P.O. Box 90698, Airport Station
Los Angeles, CA 90009

ACU
Association of Computer Users
P.O. Box 9003
Boulder, CO 80301

AEDC
American Economic Development Council
4849 North Scott Street
Suite 22
Schiller Park, IL 60176

AIA
American Institute of Architects
1735 New York Avenue, NW
Washington, DC 20006

AMA
American Management Association
135 West 50th Street
New York, NY 10020

AM/FM International
8775 East Orchard Road, Suite 820
Englewood, CO 80111

APEC Inc.
Automated Procedures for Engineering Consultants, Inc.
Miami Valley Tower
Suite 2100
40 West Fourth Street
Dayton, OH 45402

APPA
Association of Physical Plant Administrators
    of Universities & Colleges
1446 Duke Street
Alexandria, VA 22314

BOMA
Building Owners and Managers Association
    International
1250 I Street, NW
Suite 200
Washington, DC 20005

CAM-I, Inc.
611 Ryan Plaza Drive
Suite 1107
Arlington, TX 76011

CASA
Computer & Automated Systems Association
  of SME
One SME Drive
P.O. Box 930
Dearborn, MI 48121

CCCC
Coordinating Council for Computers in
  Construction
Sweet's/Dodge
1221 Avenue of the Americas
New York, NY 10020

CEPA
Society for Computer Applications in Engineering Planning and Architecture
15713 Crabbs Branch Way
Rockville, MD 20855

Council on Tall Buildings & Urban Habitat
Lehigh University/Fritz Lab
Building 13
Bethlehem, PA 18015

ICCA
Independent Computer Consultants
  Association
P.O. Box 27412
St. Louis, MO 63141

IDRC
Industrial Development Research Council
40 Technology Park–Atlanta
Norcross, GA 30092-9990

IFMA
International Facility Management
  Association

Summit Tower, Suite 1410
11 Greenway Plaza
Houston, TX 77046

IIBA
International Intelligent Buildings
  Association
1815 H Street, NW
Washington, DC 20006

Institute for Buildings and Grounds
American Public Works Association
1313 East 60th Street
Chicago, IL 60637-2881

Institute of Electrical & Electronics Engineers Computer Societies
c/o American Federation of Information Processing Societies (AFIPS)
1899 Preston White Drive
Reston, VA 22091

Institute of Industrial Engineers
25 Technology Park–Atlanta
Norcross, GA 30092

NACORE
International Association of Corporate Real
  Estate Executives
471 Spencer Drive, S
Suite 8
West Palm Beach, FL 33409

NAIOP
National Association of Industrial and Office
  Parks
1215 Jefferson Davis Highway
Suite 100
Arlington, VA 22202

NCGA
National Computer Graphics Association
2722 Merrilee Drive
Suite 200
Fairfax, VA 22031

SIGDIS
Special Interest Group on Office Information
  Systems
322 Alumni Hall
Department of Computer Sciences
University of Pittsburgh
Pittsburgh, PA 15260

SIGGRAPH
Special Interest Group for Computer
  Graphics
111 East Wacker Drive
Chicago, IL 60601

SIM
Society for Information Management
111 East Wacker Drive
Suite 600
Chicago, IL 60601

Southern Industrial Development Council
1649 Tullie Circle, NE
Atlanta, GA 30329

SUN
Space Users Network
P.O. Box 71989
Los Angeles, CA 90071-0989

ULI
Urban Land Institute
1090 Vermont Avenue, NW
Washington, DC 20005

WCGA
World Computer Graphics Association
2033 M Street, NW
Suite 399
Washington, DC 20036

Workplace Environment Group
229 Vine Street
Philadelphia, PA 19106

SHOWS AND EXPOSITIONS

ACEC Annual Convention
4849 North Scott Street
Suite 22
Schiller Park, IL 60176-9990
312/671-5646

AEDC (American Economic Development
  Council)
4849 North Scott Street
Suite 22
Schiller Park, IL 60176
312/671-5646

AIA (American Institute of Architects) National Convention & Exhibition
1735 New York Avenue, NW
Washington, DC 20006
202/626-7396

American Management Association Seminars
135 West 50th Street
New York, NY 10020
212/586-8100

ASID (American Society of Interior Designers) National Conference & Exposition
1430 Broadway, 22nd Floor
New York, NY 10018
212/944-9220

BOMA (Building Owners and Managers International) Show
1250 I Street, NW
Suite 200
Washington, DC 20005
202/289-7000

CAD
1305 Remington Road
Suite D
Schaumburg, IL 60173
312/882-0114

Computer & Management Show for the Construction Industry
425 Martingale Road
Schaumburg, IL 60173
312/240-2400

COMTEL, International Computer & Telecommunications Conferences
462 South Gilbert
Mesa, AZ 85204
602/969-7066

Constructa Hannover International Building Trade Exhibition
c/o Hannover Fairs USA, Inc.
103 Carnegie Center
Princeton, NJ 08540
609/987-1202

Council of Tall Buildings
Lehigh University/Fritz Lab
Building 13
Bethlehem, PA 18015
215/758-3515

DesCon (formerly A/E/C Systems)
Mr. George Borkovich, Principal
3400 Edge Lane
Thorndale, PA 19372
215/444-9583

Designer's Saturday, including Facilities Management Day
911 Park Avenue
New York, NY 10021
212/249-5237

IFMA (International Facility Management Association) Annual Conference & Exhibition
Summit Tower, Suite 1410
11 Greenway Plaza
Houston, Texas 77046
713/623-4362

Institute of Industrial Engineers (IIE) International Industrial Engineering Conference and Show
25 Technology Park–Atlanta
Norcross, GA 30092
404/449-0460

Intelligent Buildings Conference/Convention
International Intelligent Buildings Association
P.O. Box 11318
Newington, CT 06001
203/666-1326

International Association of Corporate Real Estate Executives (NACORE) Conference, Annual Symposium, and Exposition
471 Spencer Drive, S
Suite 8
West Palm Beach, FL 33409
305/683-8111

International Facilities Management Engineering and Operations Exposition
c/o American Institute of Plant Engineers
3975 Erie Avenue
Cincinnati, OH 45208
513/561-6000

MIT Facilities Management Conference
Office of Facilities Management Systems
Massachusetts Institute of Technology
77 Massachusetts Avenue, E19-451
Cambridge, MA 02139
617/253-6148

NAIOP (National Association of Industrial and Office Parks)
1215 Jefferson Davis Highway
Suite 100
Arlington, VA 22202
703/979-3400

NASDA (National Association of State Development Agencies) Annual Conference
444 North Capitol Street, NW
Suite 611
Washington, DC 20001
202/624-5411

NCGA (National Computer Graphics Association) Computer Graphics Show
2722 Merrilee Drive
Suite 200
Fairfax, VA 22031
703/698-9600

NCUED (National Council for Urban Economic Development) Annual Conference
1730 K Street, NW
Suite 915
Washington, DC 20006
202/223-4735

NEOCON
Merchandise Mart
Suite 470
Chicago, IL 60654
312/527-4141

OAC (Office Automation Conference)
20 Greenway Plaza
Suite 283
Houston, TX 77046
713/963-9955

Public Buildings—Today and Tomorrow
Institute for Buildings and Grounds
American Public Works Association
1313 East 60th Street
Chicago, IL 60637
312/667-2200

SIOR (Society of Industrial & Office Realtors)
777 14th Street, NW
Washington, DC 20005
202/383-1150

Tradeline (Seminars)
P.O. Box 1568
Orinda, CA 94563
415/254-1744

Urban Land Institute (ULI)
1090 Vermont Avenue, NW
Washington, DC 20005
202/289-8500

WestWeek
Pacific Design Center
8687 Melrose Avenue
Suite M-60
West Hollywood, CA 90069
213/657-0800

Workspace Conference
665 Chestnut Street
San Francisco, CA 94133
415/776-2111

APPENDIX D

PRODUCTS AND VENDORS

The following sources supply facility management products, computer systems, consulting, facility management services, and/or service bureau work.

Acme Computer
13 Spring Farm Lane
St. Paul, MN 55110

Albert C. Martin & Associates
811 West 7th Street
Los Angeles, CA 90017

Anderson Thacker Associates
1170 Westmoreland, Suite 133
El Paso, TX 79925

Apollo Computer
6800 Jericho Turnpike
Syosset, NY 11791

Applicon
32 Second Ave.
Firm Caller 514
Burlington, MA 01803-0937

Applied Software Technology, Inc.
1908 Cliff Valley Way, NE
First Floor
Atlanta, GA 30329

ARC
1824-D Fourth Street
Berkeley, CA 94710

Architects Collaborative, Inc.
639 Front Street
San Francisco, CA 94111

Arthur Andersen & Co.
33 West Monroe
Chicago, IL 60603

Arthur Young
2200 One Union Square
Seattle, WA 98101

Austin Co.
Management Consulting Division
9801 W. Higgins Road
Rosemont, IL 60018

Autosimulations
522 W. 100 North
P.O. Box 307
Bountiful, UT 84010

Auto-trol Technology Corp.
12500 North Washington Street
Denver, CO 80233

Benham Group
P.O. Box 20400
Oklahoma City, OK 73156-0400

Beyer Blinder Belle
41 East 11th Street
New York, NY 10003

Cadam Inc.
1935 North Buena Vista Street
Burbank, CA 91504

California Computer Products, Inc.
  (CalComp)
2411 West La Palma
P.O. Box 3250
Anaheim, CA 92803

California Country Trees
74-885 Jani Drive #2
Palm Desert, CA 92260

Carrier Building Services
77 West Putnam Avenue
Greenwich, CT 06830

CASA (Computer-Aided Structural Analysis)
  Gift, Inc.
7474 Greenway Center Drive
Maryland Trade Center II
Greenbelt, MD 20770

CHA Interior Architects
2500 Wilshire Blvd.
Suite 400
Los Angeles, CA 90057

Citation Manufacturing Corp.
19 South Main Street
Spring Valley, NY 10977

Cole Martinez Curtis and Associates
308 Washington Street
Marina del Rey, CA 90291

The Computer-Aided Design Group
4215 Glencoe Avenue
Marina del Rey, CA 90292

Computer-Aided Planning, Inc.
169 Monroe, NW
Grand Rapids, MI 49503

ComputerVision Corp.
100 Crosby Drive
Bedford, MA 01730

Coopers & Lybrand
1000 West Sixth Street
Los Angeles, CA 90017

CRSS, Inc.
1177 W. Loop South
Houston, TX 77027

Data General Corporation
4400 Computer Drive
Westborough, MA 01581

Data Resources, Inc.
1750 K Street, NW
Ninth Floor
Washington, DC 20006

Decision Graphics, Inc.
Two Westborough Business Park
Westborough, MA 01581

DFI/Systems
11801 Ruby Ranch Road
Orlando, FL 32819

Digital Equipment Corp.
146 Main Street
Maynard, MA 01754

Duffy Eley Giffone Worthington
8-9 Bulstrode Place, Marylebone Lane
London W1M 5FW ENGLAND

Equitable Real Estate Investment Management, Inc.
787 Seventh Avenue
New York, NY 10019

Facilities Solutions, Inc.
284 Fifth Avenue
New York, NY 10001

Facility Management Consultants
81 Main Street
White Plains, NY 10601

Facility Management Institute (FMI)
3971 South Research Park Drive
Ann Arbor, MI 48104
(Now a part of Herman Miller Research Corp.)

Facility Programmatics, Inc.
329 Cahuenga Drive
Channel Islands, CA 93030

Facility Systems Group
One Riverway, Suite 2015
Houston, TX 77056

Foresight Resources Corp.
932 Massachusetts
Lawrence, KS 66044

FORMTEC
Foster Plaza VII
661 Andersen Drive
Pittsburgh, PA 15220

Fulcrum Technologies, Inc.
560 Rochester Street
Ottawa, K1S 5K2 CANADA

Fullenwider CAD Services
400 East Rustic Road
Santa Monica, CA 90402

GE/Calma
501 Sycamore Drive
Milpitas, CA 95035

General Software Corp.
8401 Corporate Drive
Suite 500
Landover, MD 20785

Gensler & Associates, Architects
550 Kearney Street
San Francisco, CA 94108

Geo-Vision Corp.
1600 Carling Avenue, 3rd Floor
Ottawa, Ontario K12 8RS CANADA

Gibbs & Hill
11 Penn Plaza
New York, NY 10001

Graphic Horizons
125 Cambridgepark Drive
Cambridge, MA 02140

Graphic Systems, Inc.
180 Franklin Street
Cambridge, MA 02139

Haines, Lundberg, Waehler
2 Park Avenue
New York, NY 10016

Harvard Real Estate, Inc.
1350 Massachusetts Avenue
Room 1027
Cambridge, MA 02138

Hawk Systems
3633 Longview Valley Road
Sherman Oaks, CA 91403

HED Architects, Inc.
460 Seaport Court
Suite 202
Redwood City, CA 94063

Herman Miller, Inc.
8500 Byron Road
Zeeland, MI 49464

Hnedak Bobo Group
147 Jefferson Avenue
Suite 800
Memphis, TN 38103

HOK/CSC
HOK Computer Service Corp.
802 N. First Street
St. Louis, MO 63102-2529

Honeywell
2701 4th Avenue
Honeywell Plaza
Minneapolis, MN 55408

Huntington Resource
855 South Arroyo Parkway
Pasadena, CA 91105-3211

IBC
Intelligent Building Communications
220 Sansome Street
Suite 900
San Franciso, CA 94104

Interactive Systems Corp.
5500 S. Sycamore Street
Littleton, CO 80120

Intergraph Corp.
One Madison Industrial Park
Huntsville, AL 35807

Interior Acoustics, Inc.
P.O. Box 839
Bellemeade, NJ 08502

International Business Machines Corp.
400 Columbus Avenue
Valhalla, NY 10595

Jung-Brannen Research & Development
   Corp.
177 Milk Street
Boston, MA 02109

Knoll International
The Knoll Building
655 Madison Avenue
New York, NY 10012

MANCINI-DUFFY
One World Trade Center
Suite 1745
New York, NY 10048

MARKHURD Corp.
345 Pennsylvania Avenue South
Minneapolis, MN 54426

Massachusetts Institute of Technology
Office of Facilities Management Systems
77 Massachusetts Avenue E19-451
Cambridge, MA 02139

McDonnell Douglas Information Services
P.O. Box 516
St. Louis, MO 63166

MICAD Systems, Inc.
419 Park Avenue South
New York, NY 10016

Micro-Vector
One Byron Brook Place
Armonk, NY 10504

Minigraph, Inc.
620 Parkway
Broomall, PA 19008

MLA/Michael Lynn & Associates PC
300 Park Avenue South
New York, NY 10010

Monitor Facilities Management Company,
Inc.
400 Park Avenue
New York, NY 10022

Monitor Software
960 North San Antonio Road
Suite 210
Los Altos, CA 94022

Moore Productivity Software
1607 Greenwood Drive
Blacksburg, VA 24060-5937

Omni Group
Hearthstone Plaza
P.O. Box 162
Chestnut Hill, MA 02167

Parsons Brinckerhoff Inc.
250 W. 34th Street
New York, NY 10119

Peat Marwick Main & Co.
725 South Figueroa Street
Los Angeles, CA 90071

Planimetron
8700 W. Bryn Mawr Avenue
Suite 800 N
Chicago, IL 60631

Point Line Co.
2280 Powell Street
Suite 300
San Francisco, CA 94133

Price Waterhouse
400 South Hope Street
Los Angeles, CA 90071

Prime Computer
Prime Park Mail Stop 15 26
Natick, MA 01760

Pritsker & Associates, Inc.
1305 Cumberland Avenue
P.O. Box 2413
West Lafayette, IN 47906

Program Management
8950 North Central Expressway
Suite 410
Dallas, TX 75231

R. S. Means Publishing
100 Construction Plaza
Kingston, MA 02364

Ryan Group
609 Deep Valley Drive
Rolling Hills Estate, CA 90274

Sanborn, Steketee, Otis & Evans
1001 Madison Avenue
Toledo, OH 43624

SIGMA DESIGN, Inc.
61 Inverness Drive East
Suite 300
Englewood, CO 80112

Skidmore Owings & Merrill
33 West Monroe Street
Chicago, IL 60603

SKOK Systems, Inc.
222 Third Street
Cambridge, MA 02142

Steelcase, Inc.
901 44th Street, SE
Grand Rapids, MI 49508

Steinmann, Grayson, Smylie, Inc.
6310 San Vicente Boulevard
Suite 550
Los Angeles, CA 90048

Touche Ross
1633 Broadway
New York, NY 10019

Transtechnique Computer Corp.
1275 Bloomfield Avenue
Building 3/54A
Fairfield, NJ 07006

United Engineers & Constructors
30 South 17th Street
P.O. Box 8223
Philadelphia, PA 19101

University of Vermont
Technical Service Program
280 East Avenue
Burlington, VT 05401

Versacad Corp.
7372 Prince Drive
Huntington Beach, CA 92647

Walker & Associates
716 Olive Street
Los Angeles, CA 90014

Welton Becket Associates
2501 Colorado Avenue
Santa Monica, CA 90404-3585

Westinghouse Electric Corp.
2040 Ardmore Boulevard
Pittsburgh, PA 15221

Whidden/Silver, Inc.
141 West 28th Street
New York, NY 10001

Xerox Corp.
Business Systems Group
Xerox Square
Rochester, NY 14644

BIBLIOGRAPHY

BOOKS

AIA's Life Cycle Cost Analysis Task Force. 1977. *Life-Cycle Cost Analysis: A Guide for Architects.* Washington, D.C.: AIA.

Becker, F. D. 1981. *Workspace: Creating Environments in Organizations.* New York: Praeger.

Bennett, L. F. 1977. *Critical Path Precedence Networks: A Handbook on Activity-on-Node Networking for the Construction Industry.* New York: Van Nostrand Reinhold.

Brandon, P., and Moore, G. 1983. *Microcomputers in Building Appraisal.* New York: Nichols.

Computer-Aided Design Group. 1985. *An Overview of Facility Management.* Marina del Rey, Cal.: Computer-Aided Design Group.

Cross, N. 1974. *Human and Machine Roles in Computer-Aided Design.* United Kingdom: Open University, Milton Keynes.

Dell'isola, A. J., and Kirk, S. J. 1981. *Life Cycle Costing for Design Professionals.* New York: McGraw-Hill.

Eastman, C. M. 1975. *Spatial Synthesis in Computer-Aided Building Design.* London: Applied Science, and New York: Wiley.

Evans, N. *The Architect and the Computer: A Guide Through the Jungle.* London: RIBA Publications.

Evans, T. B., ed. 1984. *Facilities Management: A Manual for Plant Administration.* Washington, D.C.: Association of Physical Plant Administrators of Universities and Colleges.

Francis, R. L., and White, J. A. 1974. *An Analytical Approach.* Englewood Cliffs, N.J.: Prentice-Hall.

Gilbreath, R. D. 1983. *Managing Construction Contracts: Operational Controls for Commercial Risks.* New York: Wiley.

Glassman, S. 1981. *A Guide to Commercial Management.* Washington, D.C.: Building Owners and Managers Association International.

Gould, B. P. 1983. *Planning the New Corporate Headquarters.* New York: Wiley.

Greenberg, D.; Marcus, A.; Schmidt, A.; and Gorter, V. 1982. *The Computer Image: Applications of Computer Graphics.* Reading, Mass.: Addison-Wesley.

Hales, H. L. 1984. *Computer-Aided Facilities Planning.* New York: Marcel Dekker.

———, ed. 1985. *Computerized Facilities Planning.* Norcross, Ga.: Industrial Engineering and Management Press.

Harper, N. G., ed. 1968. *Computer Applications in Architecture and Engineering.* New York: McGraw-Hill.

Harris, D. A.; Palmer, A. E.; Lewis, M. S.; Munson, D. L.; Gershon, M.; and Gerdes, R. 1981. *Planning and Designing the Office Environment.* New York: Van Nostrand Reinhold.

Institute of Real Estate Management. 1983. *Computer Applications in Property Management Accounting.* Chicago: IREM Research Department.

Institute of Real Estate Management; National Association of Realtors; and Touche Ross & Co. 1984. *Real Estate Software Guidelines.* Chicago: IREM.

International Facility Management Association. 1984. *IFMA Report One,* Houston: IFMA.

———. 1986. *IFMA Report Two*. Houston: IFMA.

———. 1985. *Unveiling a Strategic Resource: Selected Proceedings of the Sixth Annual Conference*. Houston: IFMA.

———. 1987. *Facilities Benchmarks 1987*. Houston: IFMA.

Langue, N. *A Guide for Implementing Computer Aids in the Architectural Office*. New York: Van Nostrand Reinhold.

Mitchell, W. J. 1976. *Computer-Aided Architectural Design*. New York: Van Nostrand Reinhold.

Mitchell, W. J.; Liggett, R.; and Kvan, T. 1987. *The Art of Computer Graphics Programming*. New York: Van Nostrand Reinhold.

Molnar, J. 1983. *Facilities Management Handbook*. New York: Van Nostrand Reinhold.

Muther, R., and Hales, L. 1969. *Systematic Planning of Industrial Facilities*. Vol. 1. Kansas City: Management & Industrial Research Publications.

Negroponte, N., ed. 1975. *Reflections on Computer Aids to Design and Architecture*. New York: Petrocelli/Charter.

Newman, W. M., and Sproull, R. F. 1979. *Principles of Interactive Computer Graphics*. 2d ed. New York: McGraw-Hill.

Nora, S., and Mine, A. 1980. *The Computerization of Society*. Cambridge, Mass.: MIT Press.

Palmer, M. A. 1986. *The Architect's Guide to Facility Programming*. Washington, D.C.: American Institute of Architects.

Peña, W. M.; Caudill, W. W.; and Focke, J. W. 1977. *Problem Seeking: An Architectural Programming Primer*. Boston: Cahners Books International.

Preiser, W. F. E., ed. 1978. *Facility Programming: Methods and Applications*. Stroudsburg, Pa.: Dowden, Hutchinson & Ross.

Pulgram, W. L., and Stonis, R. E. 1984. *Designing the Office: A Guide for Architects, Interior Designers, Space Planners and Facility Managers*. New York: Watson-Guptil Publications, Whitney Library of Design.

Ralston, A., and Meek, C. L. 1982. *Encyclopaedia of Computer Science*. 2d ed. New York: Van Nostrand Reinhold.

Reynolds, R. A. 1980. *Computer Methods for Architects*. London: Butterworths.

Ripnen, K. 1974. *Office Space Administration*. New York: McGraw-Hill.

Ryan, D. L. 1979. *Computer-Aided Graphics and Design*. New York: Dekker.

Sanoff, H. 1982. *Methods of Architectural Programming*. Stroudsburg, Pa.: Dowden, Hutchinson & Ross.

Scott, J. E. 1982. *Introduction to Interactive Computer Graphics*. New York: Wiley.

Shear, M. A. 1983. *Handbook of Building Maintenance Management*. Reston, Va.: Reston Publishing.

Stitt, F. A. 1980. *Systems Drafting: Creative Reprographics for Architects and Engineers*. New York: McGraw-Hill.

Teague, L. C., Jr., and Pidgeon, C. W. 1985. *Structured Analysis Methods for Computer Information Systems*. Chicago: Science Research Associates.

Tompkins, J. A., and White, J. A. 1984. *Facilities Planning*. New York: Wiley.

Touche Ross & Co. 1984. *Real Estate Software Guidelines: Property Management*. Stroudsburg, Pa.: Institute of Real Estate Management, National Association of Realtors.

Waite, M. 1981. *Computer Graphics Primer*. Indianapolis, Ind.: Howard W. Sams.

Walker, B. S.; Gurd, J. R.; and Drawneek, E. A. 1975. *Interactive Computer Graphics*. New York: Crane Russak.

Wolfgang, K. G. 1978. *Interactive Computer Graphics*. Englewood Cliffs, N.J.: Prentice-Hall.

ARTICLES

Arden, B. W. 1980. "What Can Be Automated?" *The Computer Science and Engineering Study*. Cambridge, Mass.: MIT Press.

Area Development. 1985. "Space-Age Space Management." *Area Development*. pp. 154, 172–74, 208.

Armour, G. C., and Buffa, E. S. 1963. "A Heuristic Algorithm and Simulation Approach to Relative

Location of Facilities." *Management Science* 9(2): 294–309.

Becker, F., and Spitznagel, J. 1986. "Managing Multinational Facilities." Dallas: International Facility Management Association.

Booth, K. S., ed. 1979. "Tutorial: Computer Graphics." Long Beach, Cal.: IEEE Computer Society.

Broadbent, G., and Ward, A., eds. "An Approach to the Management of Design." *Design Methods in Architecture*. New York: Wittenborn.

Brown, D. 1986. "Terminal Solutions: How to Avoid Computer Woes." *Building Economics* (May 1986): 36–40.

Buffa, E. S.; Armour, G. C.; and Vollmann, T. E. 1964. "Allocating Facilities with CRAFT." *Harvard Business Review* 42(2): 136–59.

Campbell, D. G. 1985. "Software Aids Hard Space Use Decisions." *Los Angeles Times* (September 22, 1985).

Carstairs, E. 1985. "Can Computers Manage Space?" *Corporate Design & Realty* (March 1985).

Chasen, S. H., and Dow, J. W. 1980. "A Guide for Evaluation and Implementation of CAD/CAM Systems." *CAD/CAM Decisions* (January 1980).

Collier, L. M. 1983. "CAD-Enhanced Facilities Planning." *Design Graphics World* (July 1983): 14–17.

Dertouzos, M., and Moses, J. 1979. "The Computer Age: A Twenty-Year View." Cambridge, Mass.: MIT Press.

DeWolf, T. J. D., and Henning, N. 1980. "Office Space: Analysing Use and Estimating Needs." *Public Works Canada* (March 1980).

Duffy, F.; Colin, C.; and Worthington, J. 1976. "Planning Office Space." London: Architectural Press; New York: Nichols Publishing Co.

Eastman, C. M. 1970. "Representations for Space Planning." *Communications of the ACM* 13(4): 242–50.

———. 1970. In *Emerging Methods of Environmental Design and Planning*, ed. G. Moore. Cambridge, Mass.: MIT Press.

———. 1970. "Problem-Solving Strategies in De-

sign." In *Proceedings of the Environmental Design Research Association Conference*, ed. H. Sanoff and S. Cohn, pp. 242–50. Raleigh, N.C.: North Carolina State University School of Design.

———. 1970. "Teaching the Methods of Design: An Experiment." North Carolina State University School of Design Monograph.

———. 1972. "Preliminary Report on a System for General Space Planning." *Communications of the ACM* 15(2): 76–87.

———. 1973. "Automated Space Planning." *Artificial Intelligence* 4: 41–64.

———. 1976. "General-Purpose Building Description Systems." *Computer-Aided Design* 8(1): 209–16.

———. 1979. "The Computer as a Design Medium." Carnegie-Mellon University Department of Architecture Research Report Series.

———. 1980. "System Facilities for CAD Databases." Carnegie-Mellon University Department of Architecture Research Report Series.

———. 1982. "Recent Developments in Representation in the Science of Design." *Design Studies* 3(1): 45–52.

Eastman, C. M., and Lafue, G. 1984. "Semantic Integrity Transactions in Design Databases." Institute of Building Sciences Research Report 14, Carnegie-Mellon University.

Edwards, H. K.; Gillett, B. E.; and Hale, M. E. 1970. "Modular Allocation Technique (MAT)." *Management Science* 17(3): 161–69.

Elshafei, A. N. 1977. "Hospital Layout as a Quadratic Assignment Problem." *Operations Research Quarterly* 28(1): 167–79.

Engineering News-Record. 1985. "Data Bases Get Bigger and Better." *Engineering News-Record* (October 10, 1985).

Facility Management Institute. 1983. "Report on the Facility Management Database for the International Facility Management Association: Space and Furniture Standards." Ann Arbor, Mich.: Facility Management Institute.

Facilities Planning News. 1983. "Standards Are Catching On." *Facilities Planning News* (August 1983).

———. 1986. "IFMA Makes Education Programs a Priority in 1986." *Facility Management News* (January 1986).

Fisher, T. 1985. "Advancing Knowledge." *Progressive Architecture* (May 1985): 159–73.

———. 1985. "New Careers." *Progressive Architecture* (May 1985): 153–59.

Francis, R. L. 1967. "Sufficient Conditions for Some Optimum Property Facility Designs." *Operations Research* 15: 448–66.

Franklin, C. 1981. "How to Establish Guides for Corporate Standards." *Contract* (January 1981): 138–43.

Freeman, H., ed. 1980. "Tutorial and Selected Readings in Interactive Computer Graphics." Long Beach, Cal.: IEEE Computer Society.

Fullenwider, D. R. 1986. "Facility Management Systems." *Area Development* (May 1986): 88, 156–58.

Galle, P. 1981. "An Algorithm for the Exhaustive Generation of Building Floor Plans." *Communications of the ACM* 24(12): 813–25.

Garey, M. R., and Johnson, D. S. 1978. "Strong NP-Completeness Results: Motivation, Examples, and Implications." *Communications of the ACM* 25(3): 449–508.

Gero, J. S., and Julian, W. G. 1975. "Interaction in the Planning of Buildings." In *Spatial Synthesis in Computer-Aided Building Design*, ed. C. M. Eastman, pp. 184–229. New York: Wiley.

Gilmore, P. C. 1962. "Optimal and Suboptimal Algorithms for the Quadratic Assignment Problems." *Industrial and Applied Mathematics* 10(2): 305–13.

Goodrich, E. A. 1983. "Designs for Working: Space Standards for Space Layout, Noise, Lighting and Other Areas Provide the Foundation for the Modern Office." *Management World* (December 1983): 13–16.

Graves, G. W., and Whinston, A. 1970. "An Algorithm for the Quadratic Assignment Problem." *Management Science* 17(3): 453–71.

Hamer, J. M. 1985. "Five Warning Signs." *Leaders* (April/May/June 1985): 185.

———. 1986. "Computer-Aided Facility Management Part I: Costs and Benefits." *Area Development* (January 1986): 30, 44–45.

———. 1986. "The Efficacy of Automation in Corporate Facility Planning." *Industrial Development and Site Development Handbook* (March/April 1986): 155(2): 15.

———. 1986. "Computer-Aided Facility Management Part II: Direct and Indirect Benefits." *Area Development* (April 1986): 62, 84–87.

———. 1986. "Computer-Based Facility Planning: The Bottom Line." *Urban Land* (July 1986): 36–37.

———. 1986. "Computer-Aided Facility Management Part III: Costs and Benefits." *Area Development* (December 1986): 108–18.

Hamer, J. M., and Mitchell, W. J. 1977. "Space Planning." In *Computer Applications in Architecture*, ed. J. S. Gero. Essex, England: Applied Science Publishers.

Hamilton, I., and Burdett, J. R. F. 1982. "Computer Drafting Systems in Construction: The Buyer's Problems." *CAD82 Proceedings*. ed. A. Pipes. London: Butterworths.

Hanan, M.; Wolff, P.; and Agule, B. 1976. "Some Experimental Results on Placement Techniques." *Proceedings of the Thirteenth ACM Design Automation Conference*, pp. 214–24.

Harvey, W. L.; Hamer, J. M.; and Reeder, C. E. 1979. "The Office Planning System." Los Angeles: Morsanelli-Heumann & Associates.

Henrion, M. 1978. "Automatic Space-Planning: A Postmortem." In *Artificial Intelligence and Pattern Recognition in Computer-Aided Design*, ed. J. C. Latombe, pp. 175–96. Amsterdam: North-Holland.

Hillier, F. S. 1963. "Quantitative Tools for Plant Layout Analysis." *Journal of Industrial Engineering* 14(1): 33–40.

Hillier, F. S., and Conners, M. M. 1966. "Quadratic Assignment Problem Algorithms and the Location of Indivisible Facilities." *Management Science* 13(1): 42–57.

IFIP. 1971. "Information Distribution Aspects of

Design Methodology." In *Proceedings of the IFIP Congress*. Amsterdam: North-Holland.

Ingersoll, R. 1986. "Hype and Hypertrophy." *Design Book Review* (Winter 1986): 3.

Knell, M. 1985. "CADG + FM: Building a Better Mousetrap." *Canada's Contract* (November/December 1985): 29.

Koopmans, J. C., and Beckmann, M. J. 1967. "Assignment Problems and the Location of Economic Activities." *Econometrica* 25: 53–76.

Krauss, R. I., and Beckmann, M. J. 1968. "Computer-Aided Cost-Estimating Techniques." In *Computer Applications in Architecture and Engineering*, ed. G. N. Harper. New York: McGraw-Hill.

Kvan, T. 1983. "CADeat Emptor." *Plan and Print* (April 1983): 38–40.

———. 1983. "Plan for Computer Usage." *Sun/Coast Architect/Builder* (April 1983).

———. 1985. "Six CADD Packages." *Design Book Review* (Winter 1985).

Lamb, B. 1986. "CADD Competitors Adopt Standard Interface." *Engineering News-Record* (June 12, 1986).

Lee, R. C., and Moore, J. M. 1967. "CORELAP—Computerized Relationship Layout Planning." *Journal of Industrial Engineering* 18(3): 195–200.

Leesley, M. E.; Buchmann, A.; and Mulraney, D. 1978. "An Approach to a Largely Integrated System for Computer-Aided Design of Chemical Process Plants." In *Congress on Computers in the Development of Chemical Engineering and Industrial Chemistry* (Paris, March 1978).

Lever, P. D. 1973. "A Photomontage System for Site Planning." *Computer-Aided Design* 5(5): 103–4.

Liggett, R. S. 1972. "Floor Plan Layout by Implicit Enumeration." In *Environmental Design: Research and Practice* (Proceedings of the EDRA3/AR8 Conference), ed. W. J. Mitchell, 23-4-1–23-4-12. Los Angeles: UCLA School of Architecture and Urban Planning.

———. 1978. "An Exploration of Approximate Solution Strategies for Combinational Optimization Problems." PhD dissertation, UCLA School of Management.

———. 1980. "The Quadratic Assignment Problem: An Analysis of Applications and Solution Strategies." *Environment and Planning B* 7: 141–162.

———. 1981. "The Quadratic Assignment Problem: An Experimental Evaluation of Solution Strategies." *Management Science* 27(4): 442–57.

———. 1985. "Optimal Spatial Arrangement as a Quadratic Assignment Problem." UCLA School of Architecture and Urban Planning.

———. 1986. "Interfacing an Automated Layout Program with a CADD System." UCLA School of Architecture and Urban Planning.

Liggett, R. S., and Mitchell, W. J. 1981a. "Interactive Floor Plan Layout Technique." *Computer-Aided Design* 13(5): 289–98.

———. 1981b. "Interactive Graphic Floor Plan Layout Method." *Computer-Aided Design* 13(5): 289–97.

———. 1981c. "Optimal Space Planning in Practice." *Computer-Aided Design* 13(5): 277–88.

Los, M. 1976. "The Koopmans-Beckmann Problem: Some Computational Results." Université de Montréal, Centre de Recherche sur les Transports.

Los Angeles Business Journal. 1986. "Principles of Facility Management." *Los Angeles Business Journal* (March 17, 1986).

Lyons, P. 1982. "The Inside Story on Corporate Furniture Standards." *Facilities Design & Management* (January 1982): 65–72.

Machover, C. 1983. "An Updated Guide to Sources of Information About Computer Graphics." *IEEE Computer Graphics and Applications* (January/February 1983): 49–59.

Manufacturing Systems. 1985. "Combination Software for Facility Managers." *Manufacturing Systems* (May 1985).

Maxim, A. 1982. "Space Planning System Interactive Interface." PhD dissertation, UCLA

Graduate School of Architecture and Urban Planning.

Mitchell, W. J. 1977. "Automated Spatial Synthesis." In *Computer-Aided Architectural Design*, ch. 13. New York: Petrocelli Charter.

———. 1982. "Computer Graphics in Architectural Practice Today." *Computer Graphics World* 4.

———. 1986. "An Introduction to Computer-Aided Facility Management." Paper presented at Tokyo Conference on Facility Management.

Mitchell, W. J., and Dillon, R. L. 1972. "A Polyomino Assembly Procedure for Architectural Floor Planning." In *Environmental Design: Research and Practice* (Proceedings of the EDRA3/AR8 Conference), ed. W. J. Mitchell. Los Angeles: UCLA School of Architecture and Urban Planning.

Mitchell, W. J.; Steadman, J. P.; and Liggett, R. S. 1976. "Synthesis and Optimization of Small Rectangular Floorplans." *Environmental Planning* 3(1): 37–70.

Muther, R., and Hales, L. 1969. "Systematic Planning of Industrial Facilities." Kansas City: Management & Industrial Research Publications.

Muther, R., and McPherson, K. 1970. "Four Approaches to Computerized Layout Planning." *Journal of Industrial Engineering* (February 1970): 39–42.

Newell, A. 1969. "Heuristic Programming: Ill-Structured Problems." In *Progress in Operations Research*, vol. 3, ed. J. Aronofsky. Englewood Cliffs, N.J.: Prentice-Hall.

Newmark, H. L. 1984. "Computers for Facilities Applications III: Guide to CAD/FM Sources." *Facilities Design & Management* (September 1984): 94–103.

———. 1983. "Don't Skimp on the Interior Design Program." *Facilities Design & Management* (July/August 1983): 64–69.

———. 1986. "In Search of Facilities Management Methodologies." *Facilities Design & Management* (May 1986): .

Nugent, C. E.; Vollmann, T. E.; and Ruml, J. 1968. "An Experimental Comparison of Techniques for the Assignment of Facilities to Locations." *Operations Research* 16(1): 150–73.

Omer, A. 1980. "Structural Properties of Three-Dimensional Block Arrangements." Carnegie-Mellon University Department of Architecture Research Report Series.

Palmer, M. E. 1986. "The Current Ability of the Architecture, Engineering, and Construction Industry to Exchange CAD Data Sets Digitally." Gaithersburg, Md.: National Bureau of Standards.

Pfefferkorn, C. E. 1975. "The Design Problem Solver: A System for Designed Equipment or Furniture Layouts." In *Spatial Synthesis in Computer-Aided Building Design*, ed. C. M. Eastman, pp. 98–146.

Pooch, U. W. 1976. "Computer Graphics, Interactive Techniques and Image Processing, 1970–75: A Bibliography." *Computer* 9(8): 46–64.

Ritzman, L. P. 1972. "The Efficiency of Computer Algorithms for Plant Layout." *Management Science* 18(5): 240–48.

Sachs, R. 1983. "Thirty Office Choices (Take Your Pick)." *Office Administration and Automation* (July 1983): 34–36.

Schbipper, S. 1984. "Furniture, Space Standards Chaperone Corporate Appearance, Capital Costs." *Facilities Design & Management* (January 1984): 78–87.

Schrack, G. F. 1978. "Current Literature in Computer Graphics and Interactive Techniques." *Computer Graphics* 12(4): 114–23. [Updates appear at intervals in subsequent volumes.]

Scriabin, M., and Vergin, R. C. 1975. "Comparison of Computer Algorithms and Visual Based Methods for Plant Layout." *Management Science* 22(2): 172–81.

Seehof, J. M., and Evans, W. O. 1967. "Automated Layout Design Program." *Journal of Industrial Engineering* 18(12): 690–95.

Shaviv, E., and Gali, D. 1974. "A Model for Space Allocations in Complex Buildings: A Computer Graphic Approach." *Build International* 7: 493–517.

Slagle, J. 1964. "An Effective Algorithm for Finding Certain Minimum-Cost Procedures for Making Binary Relations." *Journal of the ACM* 2: 253–64.

Sullivan, W. G. 1986. "Models IEs Can Use to Include Strategic Non-Monetary Factors in Automation Decisions." *Journal of Industrial Engineering* (March 1986): 42, 44–50.

Teitelman, R. 1985. "Electronic Building Blocks." *Forbes* (October 1985): 153–56.

Tuchman, J., and Seltz, A. 1985. "Computer Maze Slows Managers Seeking Answers." *Engineering News-Record* (April 4, 1985).

———. 1985. "Hot New Market Lures A-E Players to Cutting Edge." *Engineering News-Record* (April 4, 1985).

Vanderburgh, L. 1983. "A Plain-Talk Guide to the Planning Process." *Facilities Design & Management* (February/March 1983): 82–85.

Vollman, T. E.; Nugent, C. E.; and Zartler, R. L. 1968. "A Computerized Model for Office Layout." *Journal of Industrial Engineering* 19(7): 321–29.

Weinzacpfel, G., and Handel, S. 1975. "Image: Computer Assistant for Architectural Design." In *Spatial Synthesis in Computer-Aided Building Design*, ed. C. M. Eastman, pp. 61–97.

Whitehead, B., and Eldarz, M. Z. 1964. "An Approach to the Optimum Layout of Single-Story Buildings." *Architects' Journal* 17: 1373–79.

Yehuda, K. 1981. "Interactive Shape Generation and Spatial Conflict Testing." Carnegie-Mellon University Department of Architecture Research Report Series.

Yehuda, K., and Eastman, C. M. 1980. "Shape Operations: An Algorithm for Spatial Set Manipulations of Solid Objects." Carnegie-Mellon University Department of Architecture Research Report Series.

INDEX

Page numbers in *italic* indicate illustrations.